T0254846

Lecture Notes in Computer Science 13168

More information about this series at https://link.springer.com/bookseries/558

Nicholas Heller · Fabian Isensee ·
Darya Trofimova · Resha Tejpaul ·
Nikolaos Papanikolopoulos ·
Christopher Weight (Eds.)

Kidney and Kidney Tumor Segmentation

MICCAI 2021 Challenge, KiTS 2021
Held in Conjunction with MICCAI 2021
Strasbourg, France, September 27, 2021
Proceedings

Editors
Nicholas Heller ⓘ
University of Minnesota
Minneapolis, MN, USA

Darya Trofimova
German Cancer Research Center (DKFZ)
Heidelberg, Germany

Nikolaos Papanikolopoulos ⓘ
University of Minnesota
Minneapolis, MN, USA

Fabian Isensee ⓘ
German Cancer Research Center (DKFZ)
Heidelberg, Germany

Resha Tejpaul
University of Minnesota
Minneapolis, MN, USA

Christopher Weight
Cleveland Clinic
Cleveland, OH, USA

ISSN 0302-9743   ISSN 1611-3349 (electronic)
Lecture Notes in Computer Science
ISBN 978-3-030-98384-0   ISBN 978-3-030-98385-7 (eBook)
https://doi.org/10.1007/978-3-030-98385-7

This Springer imprint is published by the registered company Springer Nature Switzerland AG
The registered company address is: Gewerbestrasse 11, 6330 Cham, Switzerland

Preface

This volume contains the proceedings of the second international challenge on Kidney and Kidney Tumor Segmentation (KiTS 2021), held virtually in conjunction with the International Conference on Medical Image Computing and Computer Assisted Interventions (MICCAI) in 2021. By "proceedings", we mean to say that this volume contains the papers written by participants in the challenge to describe their approach to developing a semantic segmentation approach for kidneys, kidney tumors, and kidney cysts, using the official training dataset released for this purpose, and any other publicly available datasets of their choice.

Machine learning competitions like KiTS are poised to play an ever larger role in machine learning research, especially in an application domain like medical imaging where data is so difficult to collect and even more so to release. The standardized benchmarks that competitions provide have a singular ability to elucidate which of the many proposed methods truly are superior. Given that, we believe that those of us who organize machine learning competitions have a responsibility to push the boundaries of these events to bolster their impact and rigor, while also maintaining a high level of participation.

From what we have seen, it is not often that machine learning competitions have peer reviewed proceedings. In fact, the first iteration of KiTS in 2019 did not have them, but one of the pieces of feedback that we heard after 2019 was that the impact would be greater if the participants described their approaches with more clarity and in greater detail. We thought that offering a peer-reviewed publication for contributions of sufficient quality might incentivize participants to improve this and, given the contents of this volume, we know now that we were correct. We thank the participants for their diligent efforts in putting together exceptional manuscripts to describe their approaches.

Of course, no scientific program would be successful without the huge effort put forth by the Program Committee. The Program Committee for KiTS 2021 deserves even more praise, however, because these individuals also served to provide labels for the KiTS 2021 dataset. It surely goes without saying that KiTS 2021 would not have moved forward without their tireless efforts.

November 2021

Nicholas Heller
Fabian Isensee
Darya Trofimova
Resha Tejpaul
Nikolaos Papanikolopoulos
Christopher Weight

Organization

Organizing Committee

Nicholas Heller	University of Minnesota, USA
Fabian Isensee	German Cancer Research Center, Germany
Darya Trofimova	German Cancer Research Center, Germany
Resha Tejpaul	University of Minnesota, USA
Nikolaos Papanikolopoulos	University of Minnesota, USA
Christopher Weight	Cleveland Clinic, USA

Program Committee

Susan Austin	Mayo Clinic, USA
John French	University of Missouri, USA
Ed Walczak	University of Minnesota, USA
Mark Austin	Mayo Clinic, USA
Sean McSweeney	University of Minnesota, USA
Ranveer Vasdev	University of Minnesota, USA
Sierra Virnig	Rocky Vista College of Osteopathic Medicine, USA
Chris Hornung	University of Minnesota, USA
Jamee Schoephoerster	University of Minnesota, USA
Bailey Abernathy	University of Minnesota, USA
Sarah Chan	Brigham Young University, USA
David Wu	University of Minnesota, USA
Safa Abdulkadir	University of Minnesota, USA
Ben Byun	University of Minnesota, USA
Keenan Moore	Carleton College, USA
Justice Spriggs	University of Minnesota, USA
Griffin Struyk	University of Minnesota, USA
Alexandra Austin	University of Minnesota, USA
Anna Jacobsen	University of Utah, USA
Ben Simpson	University of Minnesota, USA
Michael Hagstrom	University of Minnesota, USA
Nitin Venkatesh	University of Minnesota, USA

Contents

Automated Kidney Tumor Segmentation with Convolution
and Transformer Network 1
   Zhiqiang Shen, Hua Yang, Zhen Zhang, and Shaohua Zheng

Extraction of Kidney Anatomy Based on a 3D U-ResNet with Overlap-Tile
Strategy 13
   Jannes Adam, Niklas Agethen, Robert Bohnsack, René Finzel,
   Timo Günnemann, Lena Philipp, Marcel Plutat, Markus Rink,
   Tingting Xue, Felix Thielke, and Hans Meine

Modified nnU-Net for the MICCAI KiTS21 Challenge 22
   Lizhan Xu, Jiacheng Shi, and Zhangfu Dong

2.5D Cascaded Semantic Segmentation for Kidney Tumor Cyst 28
   Zhiwei Chen and Hanqiang Liu

Automated Machine Learning Algorithm for Kidney, Kidney Tumor,
Kidney Cyst Segmentation in Computed Tomography Scans 35
   Vivek Pawar and Bharadwaj Kss

Three Uses of One Neural Network: Automatic Segmentation of Kidney
Tumor and Cysts Based on 3D U-Net 40
   Yi Lv and Junchen Wang

Less is More: Contrast Attention Assisted U-Net for Kidney, Tumor
and Cyst Segmentations 46
   Mengran Wu and Zhiyang Liu

A Coarse-to-Fine Framework for the 2021 Kidney and Kidney Tumor
Segmentation Challenge 53
   Zhongchen Zhao, Huai Chen, and Lisheng Wang

Kidney and Kidney Tumor Segmentation Using a Two-Stage Cascade
Framework 59
   Chaonan Lin, Rongda Fu, and Shaohua Zheng

Squeeze-and-Excitation Encoder-Decoder Network for Kidney and Kidney
Tumor Segmentation in CT Images 71
   Jianhui Wen, Zhaopei Li, Zhiqiang Shen, Yaoyong Zheng,
   and Shaohua Zheng

A Two-Stage Cascaded Deep Neural Network with Multi-decoding Paths
for Kidney Tumor Segmentation   80
  Tian He, Zhen Zhang, Chenhao Pei, and Liqin Huang

Mixup Augmentation for Kidney and Kidney Tumor Segmentation   90
  Matej Gazda, Peter Bugata, Jakub Gazda, David Hubacek,
  David Jozef Hresko, and Peter Drotar

Automatic Segmentation in Abdominal CT Imaging for the KiTS21
Challenge   98
  Jimin Heo

An Ensemble of 3D U-Net Based Models for Segmentation of Kidney
and Masses in CT Scans   103
  Alex Golts, Daniel Khapun, Daniel Shats, Yoel Shoshan,
  and Flora Gilboa-Solomon

Contrast-Enhanced CT Renal Tumor Segmentation   116
  Chuda Xiao, Haseeb Hassan, and Bingding Huang

A Cascaded 3D Segmentation Model for Renal Enhanced CT Images   123
  Dan Li, Zhuo Chen, Haseeb Hassan, Weiguo Xie, and Bingding Huang

Leveraging Clinical Characteristics for Improved Deep Learning-Based
Kidney Tumor Segmentation on CT   129
  Christina B. Lund and Bas H. M. van der Velden

A Coarse-to-Fine 3D U-Net Network for Semantic Segmentation
of Kidney CT Scans   137
  Yasmeen George

3D U-Net Based Semantic Segmentation of Kidneys and Renal Masses
on Contrast-Enhanced CT   143
  Mingyang Zang, Artur Wysoczanski, Elsa Angelini, and Andrew F. Laine

Kidney and Kidney Tumor Segmentation Using Spatial and Channel
Attention Enhanced U-Net   151
  Sajan Gohil and Abhi Lad

Transfer Learning for KiTS21 Challenge   158
  Xi Yang, Jianpeng Zhang, Jing Zhang, and Yong Xia

Author Index   165

Automated Kidney Tumor Segmentation with Convolution and Transformer Network

Zhiqiang Shen[1], Hua Yang[2], Zhen Zhang[1], and Shaohua Zheng[1(✉)]

[1] College of Physics and Information Engineering, Fuzhou University,
Fuzhou, China
sunphen@fzu.edu.cn
[2] College of Photonic and Electronic Engineering,
Fujian Normal University, Fuzhou, China

Abstract. Kidney cancer is one of the most common malignancies worldwide. Early diagnosis is an effective way to reduce the mortality and automated segmentation of kidney tumor in computed tomography scans is an important way to assisted kidney cancer diagnosis. In this paper, we propose a convolution-and-transformer network (COTRNet) for end to end kidney, kidney tumor, and kidney cyst segmentation. COTRNet is an encoder-decoder architecture where the encoder and the decoder are connected by skip connections. The encoder consists of four convolution-transformer layers to learn multi-scale features which have local and global receptive fields crucial for accurate segmentation. In addition, we leverage pretrained weights and deep supervision to further improve segmentation performance. Experimental results on the 2021 kidney and kidney tumor segmentation (kits21) challenge demonstrated that our method achieved average dice of 61.6%, surface dice of 49.1%, and tumor dice of 50.52%, respectively, which ranked the 22_{th} place on the kits21 challenge.

Keywords: Convolutional neural network · Kidney tumor · Transformer

1 Introduction

Kidney cancer is one of the most common malignancies around the world leading to around 180000 deaths in 2020 [18]. Early diagnosis of kidney tumor is crucial to reduce kidney cancer mortality. Computed tomography (CT) is an effective tool for early detection and enable radiologists to study the relationship between tumor size, shape, and appearance and its prospects for treatment [11]. However, highly accurate kidney cancer diagnosis relies on the experience of doctors and the treatment subjective and imprecise. Computer-aid diagnosis (CAD) system can be used as a second observer to confirm the diagnosis and reduce the heavy burdens of radiologists.

© Springer Nature Switzerland AG 2022
N. Heller et al. (Eds.): KiTS 2021, LNCS 13168, pp. 1–12, 2022.
https://doi.org/10.1007/978-3-030-98385-7_1

Recently, deep learning-based CAD systems have been widely developed for cancer diagnosis and achieved great performance [15,22,25]. Yu et al. used the crossbar patches and the iteratively learning strategy to train two sub-models for kidney tumor segmentation [22]. Ozdemir et al. developed a CAD system for pulmonary nodule segmentation and nodule-level and patient-level malignancy classification [15]. Zheng et al. designed a symmetrical dual-channel multi-scale encoder module in the encoding layer for colorectal tumor MRI image segmentation [25]. A CAD system of kidney cancer diagnosis may include kidney and kidney tumor segmentation as well as kidney cyst segmentation. This is a challenging task because the locations, textures, shapes, and sizes of kidney tumor are diverse in CT images as shown in Fig. 1. U-Net and its variants have been widely used for end to end lesion segmentation [3,9,17,26]. U-Net is an encoder-decoder architecture where the encoder and decoder are connected by the skip connections [17]. The encoder is with stacked local operators, i.e., convolutional layers and down-sampling operators, to aggregate long-range in-formation gradually by sacrificing spatial information. The decoder is with up-sampling and convolution layers to recover spatial resolution and refine the details. The skip connections transfer the features from the encoder to the corresponding layers of the decoder, which enable information reuse.

However, U-Net has limitations to explicitly model long-range dependency be-cause the convolution are local operators. To aggregate long-range information, the encoder usually stacks several convolutional layers interlaced with down-sampling operators. Long-range dependency, i.e., large receptive field, is crucial of a model to perform accurate segmentation. Therefore, previous researches improved the U-Net to overcome this limitation implicitly by stacking more convolution layers in the blocks of U-Net. For example, MultiResUNet designed a MultiResBlock with three convolution layers to learn multi-scale information [9]. However, large amount of convolution layers stacking in a model may influence its efficiency and cause the gradient vanish by impeding the back-propagation process.

In this paper, we propose a convolution-and-transformer network (COTR-Net) for end to end kidney, kidney tumor, and kidney cyst segmentation. COTR-Net has an encoder-decoder architecture where the encoder and the decoder are connected by the skip connections. To overcome the problem mentioned above, we inserted the transformer encoder layers [19] to the encoder of COTRNet. Specifically, the encoder consists of several convolutional layers interlaced with transformer encoder layers and max-pooling operators to explicitly model long-range dependency. The decoder is composed of several up-sampling operators each of which is followed by convolution layers to recover spatial resolution and refine contexture details. In addition, we leverage the pretrained ResNet [4] to develop the encoder, which accelerates the optimization process and prevent the model from falling into local optimum. Moreover, we added the deep supervision [12] to avoid the vanishing gradient phenomenon and rapidly train the model, in which the information between the final output and the side outputs is progressively aggregated. We evaluated the proposed method on the 2021 kidney and

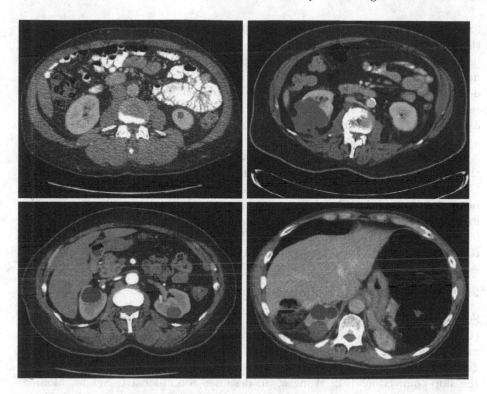

Fig. 1. Examples of an axial slice of kidney, tumor, and cyst with various locations, textures, shapes, and sizes in CT scans. kidney, tumor, and cyst are highlighted by red, green, and blue respectively. (Color figure online)

kidney tumor segmentation challenge (KITS21) [5]. Our method achieved average dice of 61.6%, surface dice of 49.1%, and tumor dice of 50.52%, respectively, and ranked the 22_{th} place on the leaderboard. Experimental results demonstrate the effectiveness of the proposed method.

2   Related Work

In the following, we review the literature related to the proposed method on two aspects including deep learning-based medical image segmentation methods and self-attention mechanism.

2.1   Medical Image Segmentation

The emergence of U-Net has greatly promoted the development of medical image segmentation [17]. Then, 3D U-Net extended the vanilla U-Net to the 3D scenario. Since then, several new networks, such as U-Net++ [26], and MultiResUNet [9] both including 2D and 3D version of architectures has been proposed for medical image segmentation by improving U-Net architecture. In kidney tumor segmentation, although CT scans has 3D spatial attribute, this task can be resolved in 2D or 3D scenario. Jackson et al. proposed an automatic segmentation framework based 3D U-Net for kidney segmentation [10]. Hou et al. designed a triple-stage self-guided network to achieve accurate kidney tumor segmentation [6]. Hu et al. presented a boundary-aware network with a shared 3D encoder, a 3D boundary decoder, and a 3D segmentation decoder for kidney and renal tumor segmentation [8].

Although processing using 3D data can reflect the whole information about the nodules, it will also require more training time and storage space. In addition, CT scans usually have different slice thicknesses, which are not recommended to be uniformly used in 3D segmentation task. On the contrary, 2D slices are not influenced by the slice thickness, and both training time and resources needed for processing are less than 3D patches. Therefore, in this work, we use 2D slice to perform the kidney tumor segmentation task.

2.2   Self-attention Mechanism

Self-attention mechanism is an effective tool for convolution neural networks (CNN) to localize the most prominent area and capture global contextual information [7,19,20]. Oktay et al. proposed an attention U-Net where the attention gates are added to the skip connections to filter the features propagated through the skip connections [14]. Wang et al. designed a non-local U-Net for biomedical image segmentation, in which the non-local block was inserted into U-Net as size-preserving processes, as well as down-sampling and up-sampling layers [21]. Zheng et al. proposed a dual-attention V-network for pulmonary lobe segmentation where a novel dual-attention module to capture global contextual information and model the semantic dependencies in spatial and channel dimensions is introduced [24]. Recently, transformer has been exploited in medical image processing [2,23]. Zhang et al. presented a two-branch architecture, which combines transformers and CNNs in a parallel style for polyp segmentation [23]. Chen et al. proposed a TransUNet in which the transformer encodes tokenized image patches from a convolution neural network (CNN) feature map as the input sequence for extracting global contexts [2]. However, these methods need large-scale GPU memory and this will not feasible for common users. Hence, we proposed a lighted transformer-based segmentation framework which needs only 8G GPU memory for network training.

3   Methods

The diagram of the proposed method is illustrated in Fig. 2. We detail the network architecture on Sect. 3.1 and the loss function on Sect. 3.2. The preprocess and postprocess methods are presented on Sect. 3.3. Other implementation details are introduced in Sect. 3.4.

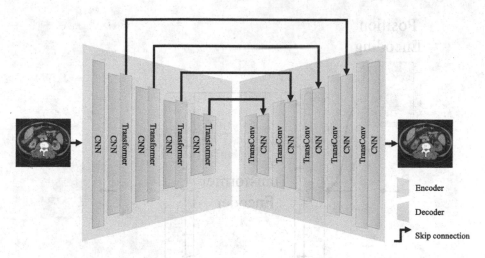

Fig. 2. The diagram of COTRNet.

3.1 Network Architecture

We proposed COTRNet for kidney and kidney tumor segmentation. The network architecture of COTRNet is shown in Fig. 2. COTRNet take slices of size 224×224 as input and output the segmentation mask having the same size as input. The motivation of the proposed COTRNet is to capture long range dependencies i.e., large receptive field, of learned features for accurate kidney and tumor segmentation. Inspired by the detection transformer (DETR) which first exploited a pretrained CNN for feature extraction and transformer for feature encoding and predictions decoding [1]. The CNN and transformer are independent with each other in DETR. Although the transformer is proficient in learning global information, it takes the sequential data as input, which disentangles spatial structure of the input images.

Instead, we integrate the transformer with CNNs where the transformer layers are inserted into the CNNs to learning long range dependencies and the CNN then recover the spatial structure of the input images. COTRNet has an UNet-like architecture, which consists of an encoder, decoder, and the skip connections. Specifically, the encoder is composed of a series of convolution layers interleaved with transformer encoder layers. The transformer encoder layer is shown in Fig. 3. An input image is first transformed to low level features by the first convolution layer, then, to features has global information by the transformer encoder layer, and finally, the next convolution layer is utilized to reconstitute the spatial structure. Through a series of operations, the output features of the encoder are the fine-grained high-level representations which will then transfer to the decoder for refinement and segmentation predictions. Besides, feature maps transferred to the decoder through the skip connections also capture the global information of the input images, which further facilitate the decoder to predict segmentation masks via feature reuse.

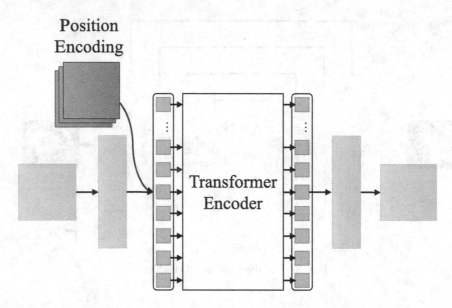

Fig. 3. The transformer encoder layer. A feature map is first transformed to a sequence, and then plus with position encoding to input into the transformer encoder layers.

To accelerate the training process and prevent the network from falling into local minimum, we leverage the pretrained parameters of ResNet18 to initialize the encoder of COTRNet. Moreover, we exploit deep supervision mechanism to promote grad-ually segmentation refinement, by supervising the hidden layers to guide training by calculating the loss of the side outputs in the intermediate stages of the decoder.

3.2 Loss Function

To overcome the data imbalance problem, we propose a class-aware weighted cross-entropy and dice (CA-WCEDCE) loss for kidney and kidney tumor segmentation. In general, the WCEDCE loss is a weighted combination of class weighted cross-entropy loss and class-weighted dice loss. The class-weighted cross-entropy (CWCE) loss is used to alleviate the inter-class unbalance problem, whereas the class-weighted dice (CWDCE) loss is exploited to solve the unbalance between each foreground class and the background class. Formally, the CA-WCEDCE loss is formulated as

$$
\begin{aligned}
\mathcal{L}_{CA-WCEDCE}(Y,\hat{Y}) &= \frac{1}{N}\sum_{i=1}^{N}\mathcal{L}_{CA-WCEDCE}\left(Y_i,\hat{Y}_i\right) = \\
&\frac{1}{N}\sum_{i=1}^{N}\left[\alpha\mathcal{L}_{CWCE}\left(Y_i,\hat{Y}_i\right) + (1-\alpha)\mathcal{L}_{CWDCE}\left(Y_i,\hat{Y}_i\right)\right]
\end{aligned}
\tag{1}
$$

where N is the batch size. α controls the contribution of the $\mathcal{L}_{CWCE}$ and $\mathcal{L}_{CWDCE}$ to the total loss $\mathcal{L}_{CAWCE}$. Y_i is the i_{th} ground truth of a batch of input images, and $\hat{Y}_i$ is the i_{th} predicted mask of a batch of predictions.

The $\mathcal{L}_{CWCE}$ is represented as

$$\mathcal{L}_{CWCE}\left(Y_i, \hat{Y}_i\right) = \frac{1}{C}\sum_{c=1}^{C} w_c 1 - \sum_{j=1}^{M} [y_{j_c} \log \hat{y}_{j_c} + (1 - y_{j_c}) \log(1 - \hat{y}_{j_c})] \quad (2)$$

And the $\mathcal{L}_{CWDCE}$ is denoted as

$$\mathcal{L}_{CWDCE}\left(Y_i, \hat{Y}_i\right) = \frac{1}{C}\sum_{c=1}^{C} w_c (1 - \frac{2\sum_{j=1}^{M} y_{jc} * \hat{y}_{jc}}{\sum_{j=1}^{M} y_{jc} + \hat{y}_{jc}}) \quad (3)$$

where C refers the total number of classes which is equal to four (kidney, tumor, cyst, and the background) in our task. M refers to the total number of pixels of the input slice in a batch. w_c denotes the weighted coefficient of the c_{th} class. y_{j_c} is j_{th} ground truth pixel of class c, and $\hat{y}_{j_c}$ is the corresponding predicted probability.

Since the deep supervision mechanism is exploited in network training, the total loss is formulated as

$$\mathcal{L} = \sum_{d=1}^{D} \beta_d \mathcal{L}_{CA-WCEDCE} \quad (4)$$

where D is the total number of output decoder. β_d is the weighted coefficient of the d_{th} decoder.

3.3 Pre- and post- processing

Preprocessing. We perform data preprocessing follows four steps.

1. Normalization. The CT scans are clipped into [-200, 300] and normalized them into [0, 255].
2. Extraction. Slices that contain foreground regions of the normalized CT scans are extracted for network training according to the ground truth masks provided by KITS21 challenge.
3. Resample. The extracted slices are resized to [224, 224] according to the input size of the pretrained model.
4. Augmentation. Data augmentation including random flip, random rotation, random crop is utilized in training process.

Postprocessing. In inference, we conducted postprocess steps as follows.

1. Transformation. The logic predictions are transformed into probabilities through the softmax function. Then, we transform the probabilistic maps to the segmentation masks according to the maximum class.
2. Resample. We resize the segmentation masks to the original size according to the raw CT scans.
3. Integration. The whole predicted mask for a raw CT scan is obtained by combining all slice segmentation masks.
4. Refinement. Morphological operations are used to refined the segmentation masks.

3.4   Implementation Details

We perform our experiments on PyTorch [16]. The models are trained via Adam optimizer with standard back-propagation with the learning rate of a fixed value of $1e-4$. We set the number of epochs as 20 and the batch size as 2. The networks are trained on a single NVIDIA GeForce GTX 1080 with 8G GPU memory.

In training, α is set as 0.5 to balance the contribution of CWCE loss and CWDCE loss. In Eq. 2, we set w_1, w_2, w_3, w_4, are set as 1, 2, 3, 4, respectively. In Eq. 4, $\beta_1 = 0.05, \beta_2 = 0.05, \beta_3 = 0.2, \beta_4 = 0.3, \beta_5 = 0.4$ to control the deep supervision mechanism. we random selected slices contained foreground objects as inputs. Data augmentation including random flip, random rotation, random crop was utilized in training process. In testing, all slices of a CT scan were input into the network to obtain predictions and the segmentation result of a CT scan was obtained by combining the predicted masks of all slices.

Table 1. Quantitative results on KITS21 training set through five-fold cross-validation. R: ResNet; T: Transformer encoder layer; P: Pretrained model; D: Deep supervision. SD: Surface Dice

Method	Kidney (Dice)	Masses (Dice)	Tumor (Dice)	Kidney (SD)	Masses (SD)	Tumor (SD)
U-Net	0.9015	0.4011	0.4148	0.8127	0.3017	0.3045
U-Net+R+T	0.9117	0.4655	0.4622	0.8598	0.3192	0.3093
U-Net+R+P+D	0.9169	0.4695	0.4838	0.8620	0.3331	0.3048
U-Net+R+T+P	0.9177	0.5082	0.5007	0.8742	0.3341	0.3435
U-Net+R+T+P+D (COTRNet)	0.9228	0.5528	0.5056	0.8853	0.3694	0.3548

4   Results

4.1   Dataset

We evaluated the proposed method on the KITS21 dataset. The KiTS21 dataset includes patients who underwent partial or radical nephrectomy for suspected renal malignancy between 2010 and 2020 at either an M Health Fairview or Cleveland Clinic medical center. KITS21 dataset includes 300 training cases of abdominal CT scans and the corresponding annotations of kidney, tumor, and cyst. The annotation files of each training case includes aggregated_AND_seg.nii.gz, aggregated_OR_seg.nii.gz, and aggregated_MAJ_seg.nii.gz. We leveraged aggregated_MAJ_seg.nii.gz in our experiments. The results of ablation study performed on the training set are shown in Sect. 4.3 and the final results on the test set are presented in Sect. 4.4.

4.2   Metrics

We used the same evaluation metrics as advocated by KiTS21 challenge, which include SØrensen-Dice and Surface Dice (SD) [13]. KITS21 leverages the hierarchical evaluation classes (HECs) to obtain a relative comprehensive measure. In an HEC, classes that are considered subsets of another class are combined with that class for the purposes of computing a metric for the superset. The HEC of kidney and masses considers kidneys, tumors, and cyst as the foreground to compute segmentation performance; the HEC of kidney mass considers both tumor and cyst as the foreground classes; the HEC of tumor considers tumor as the foreground only.

4.3   Results on KITS21 Training Set

We reported the preliminary results on KITS21 challenge training set through five-fold cross-validation. All the methods are trained with the loss function presented in Sect. 3.2. Table 1 lists the quantitative results. In general, COTR-Net outperforms other methods by a large margin, especially in tumor and cyst segmentation. Specifically, COTRNet achieved dice of 92.28%, 55.28%, and 50.56% for kidney, masses, and tumor, respectively; Measured by SD, COTRNet obtained 88.53%, 36.94%, and 35.48% for kidney, masses, and tumor, respectively. In these results, we can conclude that all components proposed in our methods contributed positively to the best performance. We also illustrate qualitative results on Fig. 4. As shown, COTRNet can accurately delineate the renal

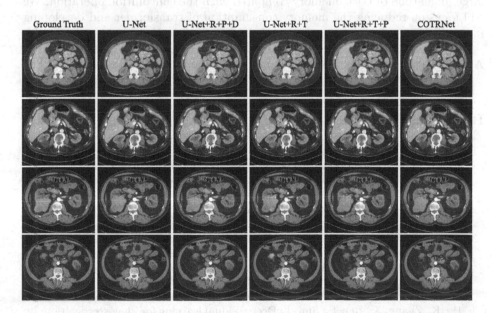

Fig. 4. Qualitative results on KITS21 dataset. R:ResNet; T:Transformer encoder layer; P:Pretrained model; D:Deep supervision.

region, tumor, and cyst. Especially in cyst segmentation, although other methods overlooked the object regions, COTRNet can correctly locate the regions and delineate the margins of the cysts.

4.4 Results on KITS21 Test Set

KITS21 test set contains 100 CT cases. Our final model were trained on the training set. Our method achieved average dice of 61.6%, surface dice of 49.1%, and tumor dice of 50.52%, respectively, which ranked the 22_{th} place on the kits21 challenge leaderboard.

5   Discussion and Conclusion

In this paper, we proposed the COTRNet to deal with kidney and tumor segmentation tasks. Inspired by the DETR that used transformer to model global information of features, COTRNet took advantage of transformer to capture long range dependencies for accurate tumor segmentation. Furthermore, we exploited pretrained parameters to accelerate convergence process. Deep supervision mechanism was used to gradually refine the segmentation results. We evaluated the proposed method on KITS21 dataset. COTRNet achieved comparable performance among kidney, cyst, and tumor segmentation. Experimental results demonstrated the effectiveness of the proposed method.

Although the transformer can explicitly model global information, it needs a large magnitude of GPU memory compared with the convolution operation. We will focus on reducing the memory consumption of transformer and developing more efficient and accurate segmentation framework in the future.

Acknowledgment. This work was supported by the Fujian Provincial Natural Science Foundation project (Grant No. 2021J02019, 2020J01472).

References

1. Carion, N., Massa, F., Synnaeve, G., Usunier, N., Kirillov, A., Zagoruyko, S.: End-to-End Object Detection with Transformers. In: Vedaldi, A., Bischof, H., Brox, T., Frahm, J.-M. (eds.) ECCV 2020. LNCS, vol. 12346, pp. 213–229. Springer, Cham (2020). https://doi.org/10.1007/978-3-030-58452-8_13
2. Chen, J., et al.: Transunet: transformers make strong encoders for medical image segmentation. arXiv preprint arXiv:2102.04306 (2021)
3. Çiçek, Ö., Abdulkadir, A., Lienkamp, S.S., Brox, T., Ronneberger, O.: 3D U-Net: learning dense volumetric segmentation from sparse annotation. In: Ourselin, S., Joskowicz, L., Sabuncu, M.R., Unal, G., Wells, W. (eds.) MICCAI 2016. LNCS, vol. 9901, pp. 424–432. Springer, Cham (2016). https://doi.org/10.1007/978-3-319-46723-8_49
4. He, K., Zhang, X., Ren, S., Sun, J.: Deep residual learning for image recognition. In: Proceedings of the IEEE Conference on Computer Vision and Pattern recognition, pp. 770–778 (2016)

5. Heller, N., et al.: The state of the art in kidney and kidney tumor segmentation in contrast-enhanced CT imaging: results of the KITS19 challenge. Med. Image Anal. **67**, 101821 (2021)
6. Hou, X., et al.: A triple-stage self-guided network for kidney tumor segmentation. In: 2020 IEEE 17th International Symposium on Biomedical Imaging (ISBI), pp. 341–344. IEEE (2020)
7. Hu, J., Shen, L., Sun, G.: Squeeze-and-excitation networks. In: Proceedings of the IEEE Conference on Computer Vision and Pattern Recognition, pp. 7132–7141 (2018)
8. Hu, S., Zhang, J., Xia, Y.: Boundary-aware network for kidney tumor segmentation. In: Liu, M., Yan, P., Lian, C., Cao, X. (eds.) MLMI 2020. LNCS, vol. 12436, pp. 189–198. Springer, Cham (2020). https://doi.org/10.1007/978-3-030-59861-7_20
9. Ibtehaz, N., Rahman, M.S.: Multiresunet: rethinking the U-Net architecture for multimodal biomedical image segmentation. Neural Netw. **121**, 74–87 (2020)
10. Jackson, P., Hardcastle, N., Dawe, N., Kron, T., Hofman, M.S., Hicks, R.J.: Deep learning renal segmentation for fully automated radiation dose estimation in unsealed source therapy. Front. Oncol. **8**, 215 (2018)
11. Kutikov, A., Uzzo, R.G.: The renal nephrometry score: a comprehensive standardized system for quantitating renal tumor size, location and depth. J. Urol. **182**(3), 844–853 (2009)
12. Lee, C.Y., Xie, S., Gallagher, P., Zhang, Z., Tu, Z.: Deeply-supervised nets. In: Artificial intelligence and statistics. In: PMLR, pp. 562–570 (2015)
13. Nikolov, S., et al.: Deep learning to achieve clinically applicable segmentation of head and neck anatomy for radiotherapy. arXiv preprint arXiv:1809.04430 (2018)
14. Oktay, O., et al.: Attention U-Net: learning where to look for the pancreas. arXiv preprint arXiv:1804.03999 (2018)
15. Ozdemir, O., Russell, R.L., Berlin, A.A.: A 3D probabilistic deep learning system for detection and diagnosis of lung cancer using low-dose CT scans. IEEE Trans. Med. Imaging **39**(5), 1419–1429 (2019)
16. Paszke, A., et al.: Pytorch: an imperative style, high-performance deep learning library. Adv. Neural Inf. Process. Syst. **32**, 8026–8037 (2019)
17. Ronneberger, O., Fischer, P., Brox, T.: U-Net: Convolutional Networks for Biomedical Image Segmentation. In: Navab, N., Hornegger, J., Wells, W.M., Frangi, A.F. (eds.) MICCAI 2015. LNCS, vol. 9351, pp. 234–241. Springer, Cham (2015). https://doi.org/10.1007/978-3-319-24574-4_28
18. Sung, H., et al.: Global cancer statistics 2020: Globocan estimates of incidence and mortality worldwide for 36 cancers in 185 countries. CA: Cancer J. Clin. **71**(3), 209–249 (2021)
19. Vaswani, A., et al.: Attention is all you need. In: Advances in Neural Information Processing Systems, pp. 5998–6008 (2017)
20. Wang, X., Girshick, R., Gupta, A., He, K.: Non-local neural networks. In: Proceedings of the IEEE Conference on Computer Vision and Pattern Recognition, pp. 7794–7803 (2018)
21. Wang, Z., Zou, N., Shen, D., Ji, S.: Non-local U-Nets for biomedical image segmentation. In: Proceedings of the AAAI Conference on Artificial Intelligence, vol. 34, pp. 6315–6322 (2020)
22. Yu, Q., Shi, Y., Sun, J., Gao, Y., Zhu, J., Dai, Y.: Crossbar-Net: a novel convolutional neural network for kidney tumor segmentation in CT images. IEEE Trans. Image Process. **28**(8), 4060–4074 (2019)

23. Zhang, Y., Liu, H., Hu, Q.: Transfuse: Fusing transformers and CNNs for medical image segmentation. arXiv preprint arXiv:2102.08005 (2021)
24. Zheng, S., et al.: A dual-attention V-network for pulmonary lobe segmentation in CT scans. IET Image Process. **15**(8), 1644–1654 (2021)
25. Zheng, S., et al.: MDCC-Net: multiscale double-channel convolution u-net framework for colorectal tumor segmentation. Comput. Biol. Med. **130**, 104183 (2021)
26. Zhou, Z., Siddiquee, R., Mahfuzur, Md., Tajbakhsh, N., Liang, J.: UNet++: a Nested U-Net architecture for medical image segmentation. In: Stoyanov, D., et al. (eds.) DLMIA/ML-CDS -2018. LNCS, vol. 11045, pp. 3–11. Springer, Cham (2018). https://doi.org/10.1007/978-3-030-00889-5_1

Extraction of Kidney Anatomy Based on a 3D U-ResNet with Overlap-Tile Strategy

Jannes Adam[1], Niklas Agethen[1], Robert Bohnsack[1], René Finzel[1],
Timo Günnemann[1], Lena Philipp[1(✉)], Marcel Plutat[1], Markus Rink[1],
Tingting Xue[1], Felix Thielke[2], and Hans Meine[1,2]

[1] University of Bremen, Bremen, Germany
len_phi@uni-bremen.de
[2] Fraunhofer MEVIS, Bremen, Germany

Abstract. In this paper we present our approach for the KiTS21 Challenge. The goal is to automatically segment kidneys, (renal) tumors and (renal) cysts based on 3D computed tomography (CT) images of the abdomen. The challenge provided public training 300 cases for this purpose. To solve this problem, we used a 3D U-ResNet with pre- and postprocessing and data augmentation. The preprocessing includes the overlap-tile strategy by preparing the input patches, while a rule-based postprocessing was applied to remove false-positive artefacts. Our model achieved 0.812 average dice, 0.694 average surface dice and 0.7 tumor dice. This led to the 12.5th position in the KiTS21 challenge.

Keywords: U-ResNet · Residual connection · Medical image segmentation

1 Introduction

Kidney cancer is a common type of cancer for which automated anatomical labeling would benefit diagnosis and treatment. In clinical practice, however, manual delineation of all relevant structures is too much effort. Therefore, the KiTS challenges have been set up to investigate automated procedures for the extraction of kidney anatomy. The KiTS21 challenge follows the previous KiTS19 challenge which already provided a dataset of labeled kidneys in CT [1]. In addition, KiTS21 provides labels for tumors and cysts. The baseline approach to this problem is a U-Net, of which the nnU-Net variation won the KiTS19 challenge. We try to improve on this baseline with a 3D U-ResNet, data augmentation, pre- and postprocessing. The residual connections of the U-ResNet should reduce the vanishing gradient problem and has also been effectively used in other medical applications [5]. In contrast to the nnU-Net, the overlap-tile strategy is used as it was originally used in the U-Net [4].

This contribution is from the "DeepAnatomy" team of master's students in computer science at the University of Bremen in collaboration with the Fraunhofer Institute for Digital Medicine MEVIS.

© Springer Nature Switzerland AG 2022
N. Heller et al. (Eds.): KiTS 2021, LNCS 13168, pp. 13–21, 2022.
https://doi.org/10.1007/978-3-030-98385-7_2

2   Methods

We trained ca. 40 variations of the U-ResNet to find better parameters and examine the effect of hyperparameters and preprocessing options. In the post-processing we do a rule based cleanup.

2.1   Training and Validation Data

Our submission made use of the official KiTS21 training set alone. We converted the annotations in majority voted labels and transversal orientation. From the 300 cases we made a randomized split of 210 training, 30 validation and 60 test cases.

2.2   Preprocessing

First the CT values are thresholded at -1000 HU. We then resample the voxel size to 1.5 mm, which we found best performing for a 3 level network. Adding more levels did not improve the performance, but due of hardware limitations we couldn't train a 5 level network.

The data then gets augmented by x-axis flipping, which swaps the kidney positions, scaling by $\pm 10\%$ and rotation. The rotation is sampled from a normal distribution with a standard deviation of $15°$. We don't use weighted inputs, but the training batches are created with a certain ratio of foreground in it. By doing this, we prevent the model from unlearning rare structures. Analyzing the KiTS21 data showed that cyst voxels are very rare. The batches are generated with a composition of patches, where 50% contain at least one voxel tumor or cyst, 25% at least one voxel cyst and 25% without constraints. Because of the patch size, the average number of cyst voxels in a batch is still smaller than 0.5%.

We separate the dataset images into smaller patches of size $32 \times 32 \times 32$ voxels, since the full 3D images do not fit onto our available hardware. The patch size of 32 voxels was chosen together with the batchsize, filter sizes, levels, etc. (described in Sect. 2.3) in order to make use of all available GPU memory and have a compromise between statistically independent samples and low overhead.

We use the overlap-tile strategy for seamless segmentation, so the architecture uses valid-mode convolutions and the input images are padded with input image context before being fed into our model to achieve an output patchsize of $32 \times 32 \times 32$. The padding size depends on the model architecture and its number of convolution and pooling layers. Our architecture (see Sect. 2.3) requires a padding of 21 voxels on both edges of every dimension, resulting in model input images of size $74 \times 74 \times 74$ voxels. The padding is implemented by cutting out $74 \times 74 \times 74$ patches from the input images (filled-up with -1000 HU outside the domain of the original CT volume) so that the output patches of size $32 \times 32 \times 32$ voxels are exactly adjacent without overlapping or gaps.

2.3 Proposed Method

To meet the challenge we decided to use a 3D U-ResNet architecture, i.e. a U-Net [4] extended by residual blocks. A residual block consists of two 3D convolution layers with kernel size of $3 \times 3 \times 3$ and strides of $1 \times 1 \times 1$, each followed by Batch Normalization and ReLu activation function (see Fig. 1).

We setup an U-ResNet with 3 levels, where every level combines a residual block, a dropout layer (with dropout rate 0.2) and another convolution layer with strides $2 \times 2 \times 2$ in the down-path (down scaling to reduce the image size) or a transposed convolution layer with strides $2 \times 2 \times 2$ in the up-path (up scaling to increase the image size). At each level the number of filters for every convolution layer of that level is doubled, while the first level starts with filters of 32. All the convolution layers of the network apply valid padding, except the layers for up- and down scaling where same padding is used. Aligned with the U-Net implementation the levels from down- and up-path of the same rank are connected via shortcuts.

The first level starts with an additional convolution layer, combined with following Batch Normalization and ReLu layer (down-path), and ends with a convolution with kernel size $1 \times 1 \times 1$ that has 4 output channels, one per output class background, kidney, tumor or cyst, followed by Softmax as final activation function (up-path).

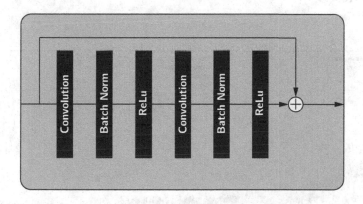

Fig. 1. Residual Block

During training batches of 15 patches are fed into the model and a dice loss function is used. As described in Sect. 2.2 we apply oversampling instead of class weighting. As optimizer we chose Adam [3] with a learning rate of 0.0001. The remaining parameters are the default ones of the tensorflow parametrization.

The validation during training is done after 800 training iterations based on 500 patches from validation data set, separated into validation batches of 50 patches.

For our validation strategy to find the best model, we built an inference network and evaluated our models using the given evaluation measures of the challenge. This includes the Sørens-Dice Coefficient and Surface Dice for the three HEC's Kidney and Masses (kidney, tumor, cyst), Kidney mass (tumor and cyst) and tumor. Also, we looked at the Dice and Surface Dice score only for the kidney and cyst alone to see how our scores come up in the HEC's. In the end, we chose the model that had the best Dice and Surface Dice scores for the three HEC's.

2.4 Postprocessing

We used connected component analysis to remove false positive fragments not connected to kidneys. This is achieved by retaining only the two components with the largest fractions of voxels labeled "kidney" (1).

Since the lesions were often not classified consistently by our model, so that there are voxels of both classes (tumor and cyst) within a single predicted lesion, an additional postprocessing step homogenizes the classes per lesion. Therefore, a connected component analysis is applied to determine all lesions from our model's output images. For each of these components, we apply a majority vote to adjust the output class of the given component to the class which occurs more frequently (number of voxels) within the given component.

3   Results

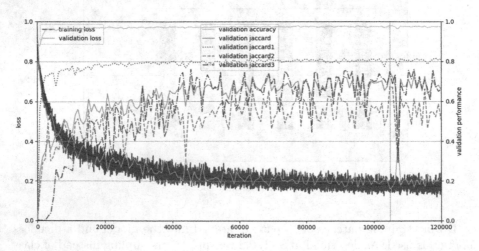

Fig. 2. Training plot. Jaccard 1 to 3 correspond to the classes kidney, tumor and cyst, respectively

Table 1. Scores on our internal test set

Dice			Surface Dice		
Kidney & Masses	Kidney Mass	Tumor	Kidney & Masses	Kidney Mass	Tumor
0.951	0.798	0.781	0.904	0.648	0.627

Dice		Surface Dice	
Kidney	Cyst	Kidney	Cyst
0.841	0.360	0.838	0.210

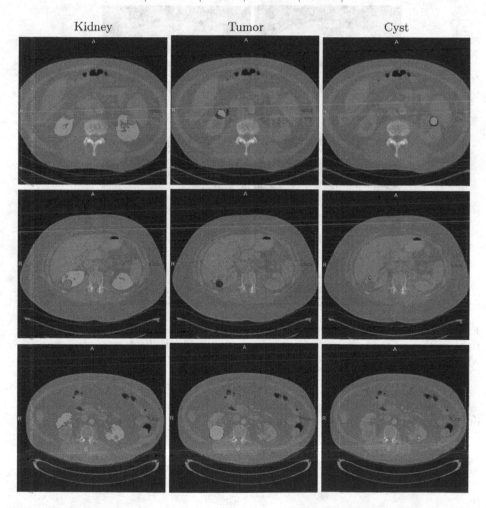

Fig. 3. Example segmentation results on three test cases, compared with the reference labels. *green:* true positive, *pink:* false positive, *blue:* false negative (Color figure online)

The training took 169601 iterations. Figure 3 shows the training progression.

In addition to the required metrics, which were measured on our test set, we also considered the Dice of kidney and cysts to better localize sources of error (see Table 1). As can be seen in Fig. 3, there are cases where the marginal regions of the cysts are incorrectly segmented as tumors or cysts and tumors are interchanged. The kidneys, on the other hand, are generally well recognized.

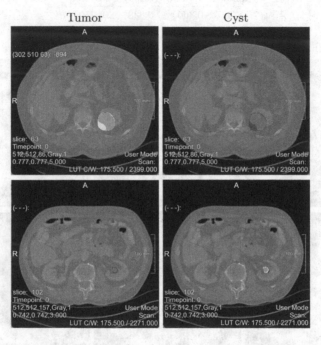

Fig. 4. Example segmentation results on two test cases, left: without post-processing, right: with post-processing. *dark green:* true positive both, *light green:* true positive with postprocessing, *dark pink:* false positive without postprocessing

The second step of post-processing corrects the false-positive classified tumor or cyst margins, as can be seen in Fig. 4.

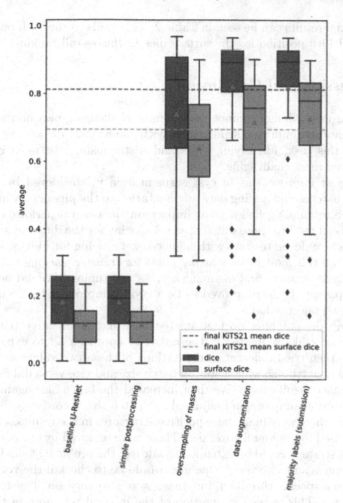

Fig. 5. Performance on average when adding different strategies

Figure 3 shows the impact of different strategies on the performance of the model. Each strategy is added to the baseline U-ResNet in addition to the previous step, e.g. data augmentation is the baseline combined with the simple postprocessing step, as well as oversampling of masses. By adding oversampling a great improvement can be observed, which is due to the improvement of cyst segmentation.

Table 2. Official results

Dice	Surface dice	Tumor dice
0.812	0.694	0.700

The official results can be seen in Table 2. This leads to the 12th position for the dice and 13th position for the surface dice in the overall ranking.

4   Discussion and Conclusion

While the segmentation of tumors was the most challenging part on the KiTS19 data, segmentation improved significantly with the KiTS21 data using the same model. For this task, identifying cysts and distinguishing between cysts and tumors proved most challenging.

The biggest improvement in cyst segmentation was achieved by oversampling in the batches and adding data augmentation to the pipeline. Additionally, changing the resampling had a great impact on the general performance. This could have been used to chose different voxel spacing for the different structures. One approach could be to start with higher voxel spacing for kidney segmentation and to use this model as a starting point for transfer learning with a lower voxel spacing to segment and distinguish cysts and tumors. We did not do this as a voxel spacing of 1.5 mm proved to be a good compromise between detailed and contextual information.

Regarding the patching we first started with maximum large patches and a batch size of 2 like suggested in the nnU-Net approach [2] to exploit GPUs memory, we then tried different configurations of these two values and noticed that increasing patch size while lowering batch size and vice versa did not impact the performance significantly. We then increased the batch size again to 15 to get a more robust sampling and adjusted the patch size accordingly.

Moreover, there were many false positive structures in the results, which were probably caused by the small voxel size. These were removed by our postprocessing, but this strategy could be extended to address the problem of small cysts on tumor structures and vice versa. One idea would be to check if the volume for a cyst is above a specific threshold, but there were also very small cysts included in the data set. This raises the question of the medical relevance of these very small cysts. In general, we can say that the close study of the data was the key to improvement.

References

1. Heller, N., et al.: The state of the art in kidney and kidney tumor segmentation in contrast-enhanced ct imaging: Results of the kits19 challenge. Med. Image Anal. **67**, 101821 (2021)
2. Isensee, F., Jaeger, P.F., Kohl, S.A.A., Petersen, J., Maier-Hein, K.H.: nnU-Net: a self-configuring method for deep learning-based biomedical image segmentation. Nat Methods **18**, 203–211 (2021). https://doi.org/10.1038/s41592-020-01008-z
3. Kingma, D.P., Ba, J.: Adam: A method for stochastic optimization (2014)

4. Ronneberger, O., Fischer, P., Brox, T.: U-Net: convolutional networks for biomedical image segmentation. In: Navab, N., Hornegger, J., Wells, W.M., Frangi, A.F. (eds.) MICCAI 2015. LNCS, vol. 9351, pp. 234–241. Springer, Cham (2015). https://doi.org/10.1007/978-3-319-24574-4_28
5. Siddique, N., Paheding, S., Elkin, C.P., Devabhaktuni, V.: U-Net and its variants for medical image segmentation: a review of theory and applications. IEEE Access **9**, 82031–82057 (2021). https://doi.org/10.1109/access.2021.3086020

Modified nnU-Net for the MICCAI KiTS21 Challenge

Lizhan Xu[✉], Jiacheng Shi, and Zhangfu Dong

Southeast University, Nanjing, China
213172291@seu.edu.cn

Abstract. KiTS21 Challenge is to develop the best system for automatic semantic segmentation of renal tumors and surrounding anatomy. The organizers provide a dataset of 300 cases and each case's CT scan is segmented to three semantic classes: Kidney, Tumor and Cyst. Compared with KiTS19 Challenge, cyst is a new semantic class, but these two tasks are quite close and that is why we choose nnUNet as our model and made some adjustments on it. Some important changes are made to the original nnUNet to adapt to this new task. Furthermore, we train models in 3 different ways and finally and merge them into one model by specific strategies. Detailed information is available in the part of Methods. The organizer uses an evaluation method called "Hierarchical Evaluation Classes" (HECs). The HEC scores of each model are showed in the following .

Keywords: Semantic segmentation · nnU-Net · Model ensemble

1 Introduction

This challenge is a semantic segmentation task of renal tumors and surrounding anatomy. The organizer provides 300 cases who undergone a contrast-enhanced preoperative CT scan that includes the entirety of all kidneys. Each case's most recent corticomedullary preoperative scan was (or will be) independently segmented three times for each instance of the following semantic classes—Kidney, Tumor, Cyst. Each instance was annotated by three independent people and final label is result of aggregating all of these files by various methods--OR, AND, MAJ. Another knowledge we use is that the scalar value of cysts in CT image is at the low level, since the cysts contain mostly water. This makes the scalar data augmentation not perform well.

Following contents in this paper:

Methods: In this part, we introduce our main approach and detailed parameters of our model.

Results: Official evaluation criteria is explained at first and then our results and training details is showed.

Discussion and Conclusion: we summarize our approach and results here, and point out the parts that can be improved.

N. Heller et al. (Eds.): KiTS 2021, LNCS 13168, pp. 22–27, 2022.
https://doi.org/10.1007/978-3-030-98385-7_3

2   Methods

We take nnU-Net as our main method and make some adjustments on it. Firstly, we adjust the data augmentation to adapt the challenge. We removed the scalar value related data augmentation strategy, as it modifies the scalar value. Secondly, we use a "two-step" method to accomplish the whole task. We put tumors and cysts into one class to train a model in the first part, and trained another model to distinguish cyst from tumor in the second part. Thirdly, we trained another model which takes kidney and cyst as a whole. At last, we take advantage of model ensemble to integrate trained models into a system. We use the "two-step" method and the traditional "one-step" three-classification method respectively, and integrate the 3 methods as our final model.

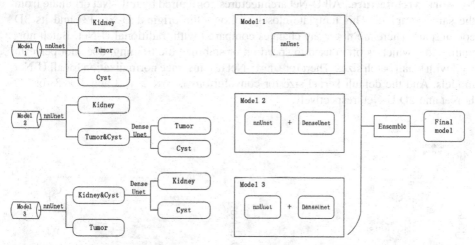

Fig. 1. The brief introduction of our main approach. Three parallel models are trained and ensembled into a final model. In fact, this simple way has a good effect on the result. DenseUNet is used here for that the lightweight net takes short time to train but has a similar effect at the same time.

2.1   Training and Validation Data

Our submission made use of the official KiTS21 training set alone.

2.2   Preprocessing

As the first processing step, nnU-Net crops the provided training cases to their nonzero region. It reduced the image size of datasets substantially and thus improved computational efficiency.

Resampling. In the medical domain, the voxel spacing (the physical space the voxels represent) is heterogeneous. To cope with this heterogeneity, nnU-Net resamples all images to the same target spacing using either third order spline, linear or nearest neighbor interpolation. In our project, the original image size is 512×512 and final resampled to 603×603.

Data Augmentation. A variety of data augmentation techniques are applied on the fly during training: rotations, scaling, Gaussian noise, Gaussian blur, brightness, simulation of low resolution, gamma and mirroring. Gray enhancement is discontinued because it changes the gray value which is important to the recognition of cysts.

ROI. It is a part only for DenseUnet. Taking advantage of the preliminary segmentation results of nnUnet, we work out the ROI (region of interest) for DenseUnet. It proves that this action will help reduce scale of DenseUnet and improve the result.

2.3 Proposed Method

Network Architecture. All U-Net architectures configured by nnU-Net originate from the same template. This template closely follows the original U-Net [3] and its 3D counterpart. There are not many changes compared with traditional U-Net. Batch normalization, which is often used to speed up or stabilize the training, does not perform well with small batch sizes. Therefore, nnU-Net use instance normalization for all U-Net models. And the default kernel size for convolutions is $3 \times 3 \times 3$ and 3×3 for 3D U-Net and 2D U-Net, respectively.

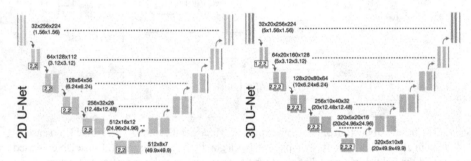

Fig. 2. The network structure of nnUnet including 3D U-Net and 2D U-Net [7].

The original author of DenseUnet [4] use it on automatic liver and tumor segmentation. DenseUnet applies the technique of densenet to U-net and has a good effect on the problem of segmentation. For each 3D input, the volume of 3D is quickly reduced to adjacent slices of 2D through the transform processing function F proposed in this paper. These 2D slices are then fed into the 2D DenseUNet to extract intra-chip features.

Loss Function. The loss function is the sum of cross-entropy and Dice loss. For each deep supervision output, a corresponding down-sampled ground truth segmentation mask is used for loss computation.

Optimization Strategy. We simply use the Adam optimizer provided by PyTorch.

Validation and Ensembling Strategy. We put each image into models trained by different methods. And to validate them, we determine the final prediction by majority voting.

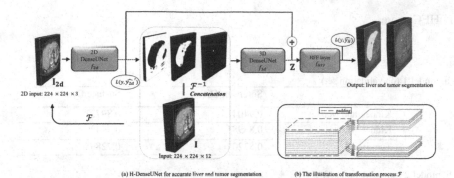

(a) H-DenseUNet for accurate liver and tumor segmentation   (b) The illustration of transformation process $\mathcal{F}$

Fig. 3. The network structure of DenseUnet [8]. The original input image size is $224 \times 224 \times 3$, in our project, the size is modified to $512 \times 512 \times 3$. And the other parameters are the same.

Post-processing. Connected component-based postprocessing [2] is commonly used in medical image segmentation. Especially in organ segmentation it often helps to remove spurious false positive detections by removing all but the largest connected component. nnUNet follows this assumption and automatically benchmarks the effect of suppressing smaller components on the cross-validation results. It can be explained by the picture below.

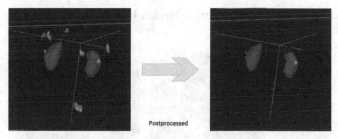

Postprocessed

Fig. 4. The demonstration of the effect of Post-processing. By taking advantage of connected component-based postprocessing, we eliminate noises around the kidney area.

3 Results

Evaluative Criteria. The organizer uses an evaluation method called "Hierarchical Evaluation Classes" (HECs). In an HEC, classes that are considered subsets of another class are combined with that class for the purposes of computing a metric for the super-set. HECs: 1. Kidney and Masses: (Kidney + Tumor + Cyst) 2. Kidney Mass: (Tumor + Cyst) 3. Tumor: (Tumor only).

Results. Here are the scores of our models and some living examples

(1) HEC scores

a. model 1 (on validation set)

	Sørensen-Dice	Surface Dice
Kidney and Masses	0.94041	0.89721
Kidney Mass	0.83983	0.74429
Tumor	0.83207	0.72894

b. model 2 (on validation set)

	Sørensen-Dice	Surface Dice
Kidney and Masses	0.93036	0.89275
Kidney Mass	0.83049	0.72003
Tumor	0.76701	0.66083

c. model 3 (on validation set)

	Sørensen-Dice	Surface Dice
Kidney and Masses	0.92845	0.89135
Kidney Mass	0.77596	0.61060
Tumor	0.79925	0.68406

(2) Some examples of predictions next to human-labels

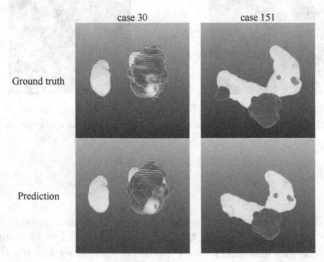

Fig. 5. Examples of case 30 and case 151.

(3) Final results from the leaderboard

Dice	Surface Dice
0.861	0.774

Training Details. Based on experience and as a trade-off between runtime and reward, all networks are trained for 1000 epochs with one epoch being defined as iteration over 250 minibatches. It took 15 days to train each model and the project took 2 months in total.

4 Discussion and Conclusion

Our project mainly takes advantage of current efficient network—nnU-Net and does not make a lot of innovation. But the power of nnU-Net is so strong and it is not easy to find a better way. There is a lot of room for improvement in our result because all our members are newbies in this field, and the time for this task is urgent. In the future, we can also try some other networks and compare the effect with the previous method.

All in all, thank you very much to the organizer for providing this opportunity. Although our result is not so perfect, we have harvested a lot in the participation.

Acknowledgements. Thank MIC-DKFZ [5] very much for nnUnet.
Thank nitsaick [6] very much for DenseUnet.

References

1. Southeast University: A College in China. Xu Lizhan, Shi Jiacheng, and Dong Zhangfu: First-Year Graduate Students in Southeast University
2. Heller, N., et al.: The state of the art in kidney and kidney tumor segmentation in contrast-enhanced CT imaging: results of the kits19 challenge. arXiv preprint. arXiv:1912.01054 (2019)
3. Ronneberger, O., Fischer, P., Brox, T.: U-Net: convolutional networks for biomedical image segmentation. In: Navab, N., Hornegger, J., Wells, W.M., Frangi, A.F. (eds.) MICCAI 2015. LNCS, vol. 9351, pp. 234–241. Springer, Cham (2015). https://doi.org/10.1007/978-3-319-24574-4_28
4. Xiaomeng, L., et al.: H-DenseUNet: hybrid densely connected UNet for liver and tumor segmentation from CT volumes. IEEE Trans. Med. Imaging 37, 12 (2018). https://doi.org/10.1109/TMI.2018.2845918
5. https://github.com/MIC-DKFZ/nnUNet
6. https://github.com/nitsaick/kits19-challenge
7. Isensee, F., Petersen, J., Kohl, S.A., Jäger, P.F., Maier-Hein, K.: nnU-Net: breaking the spell on successful medical image segmentation. ArXiv, abs/1904.08128 (2019)
8. Li, X., Chen, H., Qi, X., Dou, Q., Fu, C., Heng, P.: H-DenseUNet: hybrid densely connected UNet for liver and tumor segmentation from CT volumes. IEEE Trans. Med. Imaging 37, 2663–2674 (2018)

2.5D Cascaded Semantic Segmentation for Kidney Tumor Cyst

Zhiwei Chen and Hanqiang Liu[✉]

School of Computer Science, Shaanxi Normal University, Xi'an, China
liuhq@snnu.edu.cn

Abstract. Accurate segmentation of kidney tumours can help doctors diagnose the disease. In this work, we described a multi-stage 2.5D semantic segmentation networks to automatically segment kidney and tumor and cyst in computed tomography (CT) images. First, the kidney is pre-segmented by the first stage network ResSENormUnet; then, the kidney and the tumor and cyst are fine-segmented by the second stage network DenseTransUnet, and finally, a post-processing operation based on a 3D connected region is used for the removal of false-positive segmentation results. We evaluate this approach in the KiTS21 challenge, which shows promising performance.

Keywords: Multi-stage · Kidney and tumor and cyst segmentation · Deep learning

1 Introduction

Kidney cancer is the cancer of the genitourinary system with the highest mortality rate [1]. There are many ways to treat kidney tumors, and segmentation is only one of them. If the results of segmentation are valid, it will be helpful for subsequent tumor detection and treatment. Currently, clinical practice mainly relies on manual segmentation of the kidney and tumor, but manual segmentation brings problems such as time-consuming and laborious, and also causes inconsistent segmentation results due to differences in the subjective perceptions of physicians, which makes preoperative planning difficult, thus the need for automated segmentation is becoming more and more urgent.

In the work, motivated by the [2], we developed a two-stage neural network to locate and segment the kidney, tumor and cyst from 3D volumetric CT images. It consists of two main stages: one is rough kidney localization and the other is accurate kidney, tumor and cyst segmentation. In the first stage, only the kidney (with tumor and cyst) is segmented, and the segmentation results are stored to obtain the region of interest (ROI) and leave out the outside pixels to mitigate class imbalance and reduce memory consumption; in the second stage, we train a fine segmentation network based on the cropped kidney region obtained in the first stage. Finally, the predicted mask of the target region is transformed into a volume of the original size.

© Springer Nature Switzerland AG 2022
N. Heller et al. (Eds.): KiTS 2021, LNCS 13168, pp. 28–34, 2022.
https://doi.org/10.1007/978-3-030-98385-7_4

2 Methods

Since the kidneys make up only a small portion of the entire CT image. Each case may include non-kidney region, its segmentation is easily misled by unrelated tissues. In addition, direct segmentation of the kidney, tumor and cyst can cause difficulties in segmentation due to differences in tumor and cyst sizes and blurred boundaries between the two. This class imbalance leads to extremely difficult identification and segmentation, so the model is trained in multi- stage. Therefore, first, we train a ResSENormUNet to get a coarse segmentation of the kidney with the region of interest (ROI) in the volume. The extracted kidney then into DenseTransUNet for kidney, tumor, and cyst segmentation. As shown in Fig. 1. Finally, the results are post-processed.

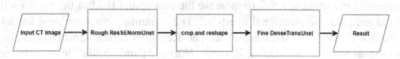

Fig. 1. The pipeline of our method

2.1 Training and Validation Data

There are 300 and 100 abdominal CT scans for training and testing in the KiTS21 Challenge dataset, respectively. Our submission made use of the official KiTS21 training set alone. We split the given 300 training CT volumes into 240 for training and 60 for validation, and evaluate the segmentation accuracy using the Dice score. For the training set, we input the original image in the first stage of the network, and in the second stage we will crop the original image and the labels directly according to the labels of the original image, and then input the second stage of the network for training. For the validation set, we input the original image for prediction in the first stage, crop the prediction result and then input it to the second stage network for further refinement.

2.2 Preprocessing

Firstly, we truncated the image intensity values of all images to the range of [−79, 304] HU to remove the irrelevant details. The choice of HU boundary values is referred to [6]. Then, truncated intensity values are normalized into the range of [0, 1] using a min-max normalization. Normalization over entire image in stage 1 benefits ROI extraction and normalization over solely ROI enhances learning targets so as to facilitate model learning [3]. Since the medical image acquisition is difficult, the amount of data is small and time-consuming to label, it is necessary to choose data enhancement for the data, which can not only add more equivalent data on the original data, but also improve the generalization ability of the model. In this paper, the data augmentation methods include horizontal flipping, random brightness contrast, random gamma, grid distortion, and flat reduction rotation, etc.

2.3 Proposed Method

2.3.1 Kidney Localization

Stage 1 of our model uses a 2.5D approach to find the ROI position of the kidney in the volume. Therefore, we merge kidney, tumor, and cyst of the target to one class. 2.5D can take more contextual information between slices into account compared to 2D, extracting more adequate features while reducing memory pressure, improving training speed compared to the 3D approach, and producing more accurate results than the 2D approach. The network structure of this stage is shown in Fig. 2, and the model is a UNet-like convolutional neural network. The model input is a stack 9 slice of adjacent axial, providing large image content in the axial plane. The model output is a segmentation map corresponding to the center slice of the stack. In the encoder, the first convolutional kernel size after the input is 7×7 to increase the perceptual field of the model without incurring significant computational overhead. The remaining convolutional kernels are all of size 3×3. The step sizes are all 2. The Rectified Linear Unit (ReLU) is used as the nonlinear activation function. Both SE Norm [4] and multiscale supervision are added. SE Norm can effectively improve the performance of the model. The SE Norm combines Squeeze-and-Excitation (SE) blocks with normalization. Similar to Instance Normalization, SE Norm layer first normalizes all channels of each example in a batch using the mean and standard deviation. Secondly, apply global average pooling (GAP) to squeeze each channel into a single descriptor. Then, two fully connected (FC) layers aim at capturing non-linear cross-channel dependencies. Besides, we apply deep supervision to enhance the discriminative ability of medium-level features. In each decoder level, we use a convolution and an upsampling to get the same spatial size as the original

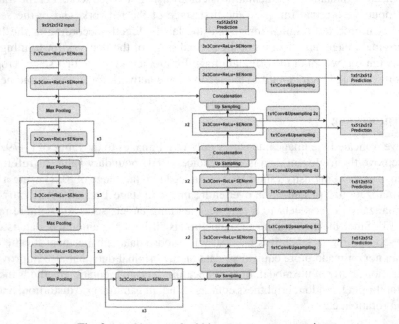

Fig. 2. Architecture for kidney coarse segmentation

image and calculate the loss using the masks from all level. The advantage is that each upsampling will be as similar as possible to the target.

2.3.2 Kidney Tumor Cyst Segmentation

The stage 2 aims at further segmentation of the kidney, tumors and cysts, and this stage uses the region of interest obtained in the previous stage as network input. The network structure of this phase is shown in Table 1, with densenet161 as the backbone network. Because adding a densely connected network is proven to enhance feature propagation, encourage feature reuse, and improve the network's ability to identify features, while mitigating the problem of gradient disappearance. In addition, adding Transformer to the model. Transformer is designed to model long-range dependencies in sequence-to-sequence tasks and capture the relations between arbitrary positions in the sequence [5]. It is powerful in modeling global context. Same as the stage 1, we also added deep supervision in DenseTransUnet.

Table 1. Architectures of the DenseTransUNet. The symbol k means kernel size, s means stride, p means padding and ch means output channels. "$() \times d$" means this block is repeated for d times.

Name	Ops	Feature map ($h \times w$)
Input	-	512×512
Convolution 1	BN + ReLU + conv($k = 7$, $s = 2$, $p = 3$, ch $= 96$)	256×256
Pooling	max pool ($k = 3$, $s = 2$, $p = 1$)	128×128
Dense block 1	$\left(\begin{smallmatrix} \text{BN+ReLU+conv(k=1,ch=192)} \\ \text{BN+ReLU+conv(k=3,p=1,ch=48)} \end{smallmatrix}\right) \times 6$	128×128
Transition layer 1	BN + ReLU + conv ($k = 1$) +average pool ($k = 2$)	64×64
Dense block 2	$\left(\begin{smallmatrix} \text{BN+ReLU+conv(k=1,ch=192)} \\ \text{BN+ReLU+conv(k=3,p=1,ch=48)} \end{smallmatrix}\right) \times 12$	64×64
Transition layer 2	BN + ReLU + conv ($k = 1$) +average pool ($k = 2$)	32×32
Dense block 3	$\left(\begin{smallmatrix} \text{BN+ReLU+ conv (k=1,ch=192)} \\ \text{BN+ReLU+ conv (k=3,p=1,ch=48)} \end{smallmatrix}\right) \times 36$ ReLU + conv ($k = 1$) + average pool ($k = 2$)	32×32
Transition layer 3	BN + ReLU + conv ($k = 1$) + average pool ($k = 2$)	16×16
Dense block 4	$\left(\begin{smallmatrix} \text{BN+ReLU+conv (k=1,ch=192)} \\ \text{BN+ ReLU + conv (k=3,p=1,ch=48)} \end{smallmatrix}\right) \times 24$	16×16
Transformer layer	conv ($k = 3$, ch $= 2208$, $s = 1$, $p = 1$) + reshape +Multi-Head Attention (MHA) block +Feed Forward Network (FFN) + reshape	1×256
Upsampling layer 1	transposed conv ($k = 3$, $s = 2$, $p = 1$) +skip connection (dense block 3) +conv($k = 3$, $p = 1$, ch $= 768$) + BN + ReLU	32×32

(continued)

Table 1. (*continued*)

Name	Ops	Feature map ($h \times w$)
Upsampling layer 2	transposed conv (k = 3, s = 2, p = 1) +skip connection (dense block 2) + conv (k = 3, p = 1, ch = 384) + BN + ReLU	64 × 64
Upsampling layer 3	transposed conv (k = 3, s = 2, p = 1) + skip connection (dense block 1) + conv (k = 3, p = 1, ch = 96) + BN + ReLU	128 × 128
Upsampling layer 4	transposed conv (k = 3, s = 2, p = 1) +skip connection (convolution 1) + conv (k = 3, p = 1, ch = 96) + BN + ReLU	256 × 256
Upsampling layer 5	transposed conv (k = 3, s = 2, p = 1) +conv(k = 3, p = 1, ch = 96) + BN + ReLU	512 × 512
Convolution 2	k = 3, p = 1, ch = 4	512 × 512

2.3.3 Post-processing

A post-processing method based on three-dimensional connectivity domain analysis was used to calculate the area of the region consisting of each detected marker object, leaving only the portion of the region area larger than the threshold as the maximum connected region of the kidney with or without cancerous tissue. Since the tumor will connect with the kidney and given by the prior knowledge that no more than two kidneys exist in the abdomen. Therefore, first, we merge the kidneys and tumor of the segmentation result and ignore the background. If this component is smaller than the second largest component multiplied by 0.8, we remove it. Second, we perform another post-processing based on the connected region only for the tumor. We will remove if this component is smaller than the largest component multiplied by 0.4.

2.3.4 Loss and Optimization

All networks are trained with stochastic gradient descent and a batch size of 16. The unweighted sum of the Generalized Dice Loss and the Focal Loss is utilized to train the model. Use Adam optimizer with initial learning rate of 1e−4 and multiplied by 0.1 when loss is not decrement in 5 epochs. Each model was trained for 100 epochs.

3 Results

We use the Dice coefficient, which is widely used in medical image segmentation, to quantitatively evaluate the accuracy of the model. An example of our prediction results is depicted in Fig. 3.

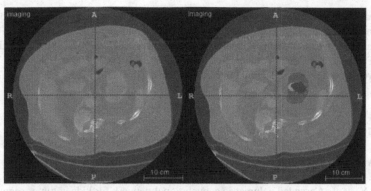

Fig. 3. An example of prediction results of case 256. The kidney is shown in red, the tumor in green, and the cyst in blue.

Table 2 shows the results of our model on the validation dataset. We validated our method on the KiTS21 challenge, the performance is shown in Table 3. Training was done on Nvidia GeForce RTX 3090 GPU (single GPU training). All networks were implemented with the PyTorch framework. It took about 5 days for training the model in the first stage and about 2 days for it in the second stage.

Table 2. The experimental results in validation data using our method.

	Kidney	Tumor	Cyst
Average dice on validation dataset	0.9430	0.7779	0.7099

Table 3. Results on the KiTS21 challenge test set.

	Mean sampled average dice	Mean sampled average SD	Position
Ours	0.8462	0.7454	10

4 Discussion and Conclusion

In this paper, we propose a segmentation method based on a multi-stage stepwise refinement approach for the segmentation of kidney, tumor and cyst in abdominal enhanced CT images. A 2.5D approach is used for data input in network training to preserve certain contextual semantic information while relieving memory pressure. In addition, this paper adopts a post-processing method based on the 3D connected domain to remove the false positive regions in the segmentation results and further improve the segmentation accuracy. There are issues in this paper that need further study and the network and methods can be improved for smaller kidneys, tumor and cyst segmentation.

Acknowledgements. This work is supported by the National Natural Science Foundation of China (Grant Nos. 62071379, 62071378 and 61571361), the Natural Science Basic Research Plan in Shaanxi Province of China (Grant Nos. 2021JM-461 and 2020JM-299), the Fundamental Research Funds for the Central Universities (Grant No. GK201903092), and New Star Team of Xi'an University of Posts & Telecommunications (Grant No. xyt2016-01).

References

1. Gelb, A.B.: Renal cell carcinoma. Cancer **80**(5), 981–986 (2015)
2. Tsai, Y.-C., Sun, Y.-N.: KiTS19 Challenge Segmentation (2019)
3. Zhang, Y., et al.: Cascaded volumetric convolutional network for kidney tumor segmentation from CT volumes. ArXiv abs/1910.02235 (2019)
4. Iantsen, A., Visvikis, D., Hatt, M.: Squeeze-and-excitation normalization for automated delineation of head and neck primary tumors in combined PET and CT images. In: Andrearczyk, V., Oreiller, V., Depeursinge, A. (eds.) HECKTOR 2020. LNCS, vol. 12603, pp. 37–43. Springer, Cham (2021). https://doi.org/10.1007/978-3-030-67194-5_4
5. Wang, W., Chen, C., Ding, M., Li, J., Yu, H., Zha. S.: TransBTS: multimodal brain tumor segmentation using transformer. arXiv:2103.04430 (2021)
6. Isensee, F., Maier-Hein, K.H.: An attempt at beating the 3D U-Net. arXiv preprint arXiv:1908.02182 (2019)

Automated Machine Learning Algorithm for Kidney, Kidney Tumor, Kidney Cyst Segmentation in Computed Tomography Scans

Vivek Pawar[✉] and Bharadwaj Kss

Endimension Technology Private Limited (Incubator Under SINE IIT Mumbai), Mumbai, India
vivek_pawar@endimension.com

Abstract. In this paper, we have described an automated algorithm for accurate segmentation of kidney, kidney tumors, and kidney cysts from CT scans. The Dataset for this problem was made available online as part of KiTS21 Challenge. Our approach was placed 13th in the official leaderboard of the competition. Our model uses a 2 stage Residual Unet architecture. The first network is designed to predict (Kidney + Tumor + Cyst) regions. The second network predicts segmented tumor and cyst regions from the output of the first network. The paper contains implementation details along with results on the official test and internal set.

Keywords: KiTS21 · Kidney segmentation · Unet

1 Introduction

There are more than 400,000 new cases of kidney cancer each year [1], and surgery is its most common treatment [2]. KiTS21 challenge [3] was conducted to accelerate the development of reliable kidney and kidney tumor semantic segmentation methodologies. Ground truth semantic segmentation for abdominal CT scans of 300 unique kidney cancer patients were provided as part of the training dataset for the challenge. The submission models are then evaluated on a test set of 45 patients (part of 300 CT scans) which is separate from the official test set of 100 cases.

2 Methods

2.1 Training and Validation Data

Our submission made use of the official KiTS21 training set alone. We divided the official KiTS21 dataset into training, validation, and internal test set. We used a validation set to finetune our approach. The approach which had the best score on the internal test set was used to create the final submission. The internal test and validation set are sampled from 300 cases, initially provided in the KiTS21 challenge.

The below table shows the distribution of scans among different sets (Table 1).

© Springer Nature Switzerland AG 2022
N. Heller et al. (Eds.): KiTS 2021, LNCS 13168, pp. 35–39, 2022.
https://doi.org/10.1007/978-3-030-98385-7_5

Table 1. Distribution of available samples between training, validation, and test set

Training	Validation	Internal test	Official test set
225	30	45	100

2.2 Preprocessing

We use a two-stage segmentation process, the first stage for Kidney + Tumor + Cyst segmentation. For the first stage, we resample all cases to a common voxel spacing of $3.22 \times 1.62 \times 1.62$ mm, with a patch size of $80 \times 160 \times 160$, for the training cases. The voxel spacing $3.22 \times 1.62 \times 1.62$ mm was chosen on the basis of the median spacing of the training dataset which is $3.22 \times 0.81 \times 0.81$ mm. The x and y spacing was doubled to avoid training large numbers of patches. Before creating the patches for training we cropped the abdomen region from the entire CT scan to avoid training of unnecessary negative patches. We used bicubic interpolation to resample the cases.

Each case is then clipped to the range $[-80, 304]$. We then subtract 101 and divide by 76 to bring the intensity values in a range that is more easily processed by CNNs. The clipping range was selected after analyzing the voxels covered by kidney, tumor, and cyst segmentation in the CT scans over the entire training set. The -80 and 304 are values chosen based on the distribution. Similarly, 101 and 76 are the mean and standard deviation of the distribution.

For the Second Stage, we resampled the voxel spacing to $0.78 \times 0.78 \times 0.78$ mm and used the patch size of $64 \times 128 \times 128$. Since we had Kidney + Tumor + Cyst Segmentation output from the first stage, we only used the patches where the first network predicted any region. Similar to the first stage, We clipped voxels to the range between $[-31, 208]$, then subtract by 55 and divide by 65 based on the voxel covered by Tumor and Cyst segmentation in CT scans.

2.3 Network Architecture

For both stages, we used base 3D unet architecture with residual blocks. Both Networks use 3D convolutions, ReLu nonlinearities, and batch normalization. We double the number of feature maps with each DownBlock in the unet. We perform downsampling till the feature map size reaches 4. For the Upsampling block, we used the Scale of 2 to upsample input. The first stage takes a single channel input of shape $80 \times 160 \times 160$ and outputs the tensor of the same shape which represents Kidney + Tumor + Cyst mask as a single entity. Similarly, the second network gives 2 channel outputs representing tumor and cyst regions respectively (Fig. 1).

This architecture of both networks uses residual blocks instead of simple convolution sequences, which is implemented in a similar fashion as conv-batchnorm-relu-conv-batchnorm-relu. The addition of the residual takes place before the last relu.

2.4 Network Training

All the stages are trained with Adam optimizer and an initial learning rate of 1e−3. We reduce the learning rate by a factor of 5. We stopped the training after 50 epochs with stage

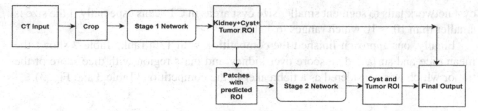

Fig. 1. Overview of the two-stage segmentation process

1 taking 80 minutes per epoch to train and stage 2 with 75 minutes per epoch. The batch sizes for stage 1 and stage 2 are 2 and 4 respectively. We used flipping, rotation, scaling, brightness, contrast, and gamma augmentations to augment patches during training. Loss function was a combination of cross-entropy and dice loss. The training was done on Nvidia GeForce GTX 1080 Ti GPUs. All networks were implemented with the PyTorch framework (Fig. 2).

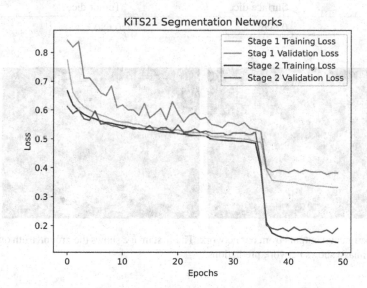

Fig. 2. Training and validation loss for stages 1 and 2

3   Results

We did not experiment with different network architectures. We tested our networks on internal test data consisting of 45 samples. The evaluation metric uses the same setup as challenge evaluation. It consists of two metrics Sørensen-Dice and Surface Dice [4]. The score for the kidney is computed by treating all the labels i.e. kidney, tumor, and cyst labels as foreground and rest as background. Similarly Mass consists of labels from tumor and cyst. Our network is very effective at detecting both kidney and tumor in most cases. However, the segmentation of cysts needs improvement. We noticed that the cyst network fails to segment smaller size cyst areas in CT scans especially if the size is smaller than 10×10 which ranges to 4 slices.

Finally, our approach finished the competition with 13th rank. Table 2 shows the mean dice and surface dice score over kidney and mass region with dice score of the tumor which was considered as a tiebreaker in the competition (Table 3 and Fig. 3).

Table 2. Experiment results on the internal test set

	Kidney	Mass	Tumor	Mean
Dice	0.96	0.82	0.77	0.85
Surface dice	0.93	0.71	0.67	0.77

Table 3. Results on the official test set

Dice	Surface dice	Tumor dice
0.784	0.696	0.680

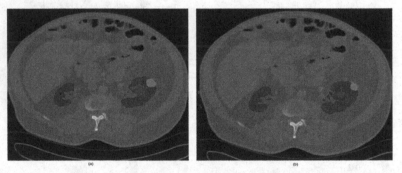

Fig. 3. Experimental outputs from our network. The first image shows the ground truth on a slice. The second image shows network predictions.

4   Discussion and Conclusion

In the preprocessing section of the paper, we mentioned that we used the same clipping values as the second stage which has tumor and cyst voxel values. Surprisingly using values based on voxels covering cyst does not yield better results for cyst segmentation. Due deadline of the competition and time-consuming training methods we did not explore the effects of other Unet variations with respect to results on the data.

In conclusion, We described a two-stage semantic segmentation pipeline for kidney and tumor segmentation from 3D CT images. Final evaluation results on KiTS21 challenge results are 0.78, 0.69, and 0.68 average dice, average surface dice, and dice score of tumor respectively placing our approach to the 13th rank on the competition leaderboard.

References

1. Kidney Cancer Statistics. wcrf.org/dietandcancer/cancer-trends/kidney-cancer-statistics
2. Cancer Diagnosis and Treatment Statistics. Stages — Mesothelioma — CancerResearch UK, 26 October 2017
3. KiTS21 Challenge. https://kits21.kits-challenge.org/
4. Nikolov, S., et al.: Deep learning to achieve clinically applicable segmentation of head and neck anatomy for radiotherapy. arXiv preprint arXiv:1809.04430 (2018)

Three Uses of One Neural Network: Automatic Segmentation of Kidney Tumor and Cysts Based on 3D U-Net

Yi Lv[1,2] and Junchen Wang[2,3(✉)]

[1] North China Research Institute of Electro-Optics, Beijing 100015, China
[2] School of Mechanical Engineering and Automation, Beihang University, Beijing 100191, China
wangjunchen@buaa.edu.cn
[3] Beijing Advanced Innovation Center for Biomedical Engineering, Beihang University, Beijing, China

Abstract. Medical image processing plays an increasingly important role in clinical diagnosis and treatment. Using the results of kidney CT image segmentation for three-dimensional reconstruction is an intuitive and accurate method for diagnosis. In this paper, we propose a three-step automatic segmentation method for kidney, tumors and cysts, including roughly segmenting the kidney and tumor from low-resolution CT, locating each kidney and fine segmenting the kidney, and finally extracting the tumor and cyst from the segmented kidney. The results show that the average dice of our method for kidney, tumor and cysts is about 0.93, 0.57, 0.73.

Keywords: Medical image segmentation · Deep learning · Neural network

1 Introduction

Segmentation and reconstruction of CT or MRI medical images is the main source of navigation data, i.e. anatomical structure of tissues and organs [1–4]. At present, the most commonly used method of medical image segmentation is still manual segmentation, which takes a long time and depends on the operator's experience and skills [5]. In recent years, breakthroughs have been made in the research of neural networks [6–9]. The deep learning technology based on neural networks can achieve fast segmentation, and effectively solve the problem of low accuracy and long time-consuming image segmentation [10]. In the field of medical image segmentation, the breakthrough of in-depth learning began with the introduction of Full Convolutional Neural Network (FCN), and another breakthrough of neural network architecture U-Net made it possible to achieve high-precision automatic segmentation of medical images. Long Jonathan et al. [11] proposed Fully Convolutional Networks structure in 2015. In the same year, Olaf Ronneberger et al. [12] proposed the U-Net network structure. U-Net is a semantic

© Springer Nature Switzerland AG 2022
N. Heller et al. (Eds.): KiTS 2021, LNCS 13168, pp. 40–45, 2022.
https://doi.org/10.1007/978-3-030-98385-7_6

segmentation network based on FCN, which is suitable for medical image segmentation. With the proposal of 3D convolutional neural networks such as 3D U-Net [13] and V-Net [14], the segmentation accuracy of some organs has reached a milestone. For example, in the MICCAI challenge 2019 kits19 competition, the accuracy of 3D U-Net in the task of kidney segmentation is very close to that of human, but the required time to complete a segmentation is far less than that of manual segmentation. The deep learning-based methods not only surpass the traditional algorithms, but also approach the accuracy of manual segmentation.

In this paper, we propose a three-step automatic segmentation method for kidney, tumors and cysts, including roughly segmenting the kidney and tumor from low-resolution CT, locating each kidney and fine segmenting the kidney, and finally extracting the tumor and cyst from the segmented kidney.

2   Methods

2.1   Network Architecture

The network architecture (3D U-Net) is illustrated in Fig. 1. We choose 3D U-Net as the neural network for the all three steps. 3D U-net includes an encoding path and a decoding path, each of which has four resolution levels. Each layer of the encoding path contains two $3 \times 3 \times 3$ convolution, each followed by a ReLu layer, followed by a $2 \times 2 \times$ Maximum pool layer with step size of 2 in each direction of 2. In the decoding path, each layer contains a 2 with a step size of $2 \times 2 \times 2$, followed by two $3 \times 3 \times 3$, each followed by a RuLu layer.

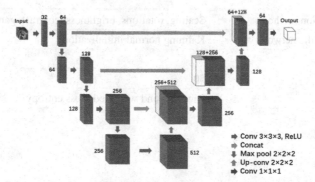

Fig. 1. The network architecture. Blue cuboids represent feature maps. The number of channels is denoted next to the feature map. (Color figure online)

2.2   Segmentation from Low-Resolution CT

Considering that the complete CT sequence is too large for the GPU, we scaled the spacing of all CT sequences to 3.41, 1.71 and 1.71. Then we cut the size of the sequence to 128 * 248 * 248 under the condition of constant spacing. Finally, we uniformly scale all the sequence sizes to 96 * 128 * 128 for the training of neural network. Then we use the generated data for training of the 3D U-Net for the first step.

2.3   Fine Segmentation of Kidney

According to the results of rough segmentation in the previous step, we first locate the position of two (or one) kidneys. Secondly, we crop the CT image with the region located and scale the size to 80 * 128 * 128. Then, we set the kidney, tumor and cyst to a unique label as the training data for the second step. Finally, we conduct the data augmentation including random translation, scaling, and trained the second 3D U-Net with the generated data.

2.4   Segmentation of Tumor and Cysts

To prevent the network from identifying areas outside the kidney as tumors or cysts, only the region of kidney segmented in the above step is used as the input for the network in step 3. Tumor and cysts were labeled with 1 and 2, respectively. After segmentation, we scale the segmented results to the original size as the final output.

2.5   Training Protocols

All the algorithms were implemented using Pytorch1.2 with Python3.7 and ran on a workstation with a AMD 5800X CPU, 32G memory and a NVIDIA RTX8000 GPU. The training protocols of the proposed method is shown in Table 1.

Table 1. Training protocols

Data augmentation methods	Scaling, rotations, brightness, contrast, gamma
Initialization of the network	Kaiming normal initialization
Batch size	4
Total epochs	50
Loss function	Dice loss and weighted cross entropy

3   Results

As the accuracy metrics, the Dice similarity coefficient (DSC), average symmetric surface distance (ASSD [mm]) and surface distance deviation (SDD [mm]) between the predicted mask and the ground truth mask were employed. Assume A and B are two masks, these metrics are given by (1), (2) and (3), where S(A) and S(B) are the surface points of A and B, respectively, and d(a, S(B)) is the minimum Euclidian distance between the point a and the points on the surface S(B). We perform 5-fold cross-validation on 300 data sets, but due to limited time, only one-fold has finished at the time of submission. The result is shown in Table 2. Figure 2 shows the results with voxel-based rendering from three examples in the evaluation dataset. Figure 3 shows the segmentation results on CT slices.

$$DSC = \frac{2(A \cap B)}{A + B} \tag{1}$$

$$ASSD = \frac{1}{|S(A)| + |S(B)|} \times \left(\sum_{a \in S(A)} d(a, S(B)) + \sum_{b \in S(B)} d(b, S(A)) \right) \tag{2}$$

$$SDD = \sqrt{\frac{1}{|S(A)| + |S(B)|} \times \left(\sum_{a \in S(A)} (d(a, S(B)) - ASSD)^2 + \sum_{b \in S(B)} (d(b, S(A)) - ASSD)^2 \right)} \tag{3}$$

Table 2. Dice comparison on three structures

	Dice	ASSD (mm)	SSD (mm)
Kidney	0.93	1.24	2.45
Tumor	0.57	8.24	10.86
Cysts	0.73	6.70	6.90

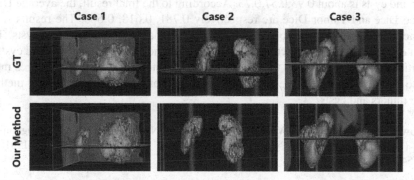

Fig. 2. Segmentation results with voxel-based rendering from three examples in the evaluation dataset. For each example, the ground truth and the segmentation results are given for visual comparison. (yellow: cysts, red: kidney, brown: tumor) (Color figure online)

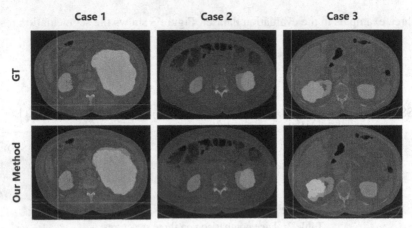

Fig. 3. Segmentation results on CT slices. The two rows are the ground truth and segmentation results of our method. (yellow: cysts, red: kidney, brown: tumor) (Color figure online)

4   Discussion and Conclusion

We propose a three-step automatic segmentation method for kidney, tumor and cysts based on 3D U-Net. The results show that the average dice of our method for kidney, tumor and cysts is about 0.93,0.57,0.73. According to the final result, the average Dice, Surface Dice and Tumor Dice are respectively 0.781, 0.618, 0.660. The results show that the accuracy of our method on kidney is better than that of tumor and cysts. The region of the kidney can be accurately identified, but the accuracy of tumor and cysts is not satisfactory. As limited by the competition time, the neural network requires more time to be fully trained. Future work will focus on promoting accuracy of our method on the tumors and cysts.

Acknowledgements. Junchen Wang was funded by National Key R&D Program of China (2017YFB1303004) and National Natural Science Foundation of China (61701014, 61911540075).

References

1. Couteaux, V., et al.: Kidney cortex segmentation in 2D CT with U-Nets ensemble aggregation. Diagn. Interv. Imaging **100**(4), 211–217 (2019)
2. Yang, Y., Jiang, H., Sun, Q.: A multiorgan segmentation model for CT volumes via full convolution-deconvolution network. Biomed Res. Int. **2017**, pp. 1–9 (2017)
3. Fu, Y., et al.: A novel MRI segmentation method using CNN-based correction network for MRI-guided adaptive radiotherapy. Med. Phys. **45**(11), 5129–5137 (2018)
4. Zhao, C., Carass, A., Lee, J., He, Y., Prince, J.L.: Whole brain segmentation and labeling from CT using synthetic MR images. In: Wang, Q., Shi, Y., Suk, H.-I., Suzuki, K. (eds.) MLMI 2017. LNCS, vol. 10541, pp. 291–298. Springer, Cham (2017). https://doi.org/10.1007/978-3-319-67389-9_34

5. Li, J., Zhu, S.A., Bin, H.: Medical image segmentation techniques, Sheng wu yi xue gong cheng xue za zhi = Journal of biomedical engineering = Shengwu yixue gongchengxue zazhi **23**(4), 891–894 (2006)
6. Micheli-Tzanakou, E.: Artificial neural networks: an overview. Netw.-Comput. Neural Syst. **22**(1–4), 208–230 (2011)
7. LeCun, Y., Bengio, Y., Hinton, G.: Deep learning. Nature **521**(7553), 436–444 (2015)
8. Zhang, J., Zong, C.: Deep neural networks in machine translation: an overview. IEEE Intell. Syst. **30**(5), 16–25 (2015)
9. Zarandy, Á., Rekeczky, C., Szolgay, P., Chua, L.O.: Overview of CNN research: 25 years history and the current trends. In: 2015 IEEE International Symposium on Circuits and Systems, pp. 401–404. IEEE (2015)
10. Kim, D.Y., Park, J.W.: Computer-aided detection of kidney tumor on abdominal computed tomography scans. Acta Radiol. **45**(7), 791–795 (2004)
11. Shelhamer, E., Long, J., Darrell, T.: Fully convolutional networks for semantic segmentation. IEEE Trans. Pattern Anal. Mach. Intell. **39**(4), 640–651 (2017)
12. Ronneberger, O., Fischer, P., Brox, T.: U-Net: convolutional networks for biomedical image segmentation. In: Navab, N., Hornegger, J., Wells, W.M., Frangi, A.F. (eds.) MICCAI 2015. LNCS, vol. 9351, pp. 234–241. Springer, Cham (2015). https://doi.org/10.1007/978-3-319-24574-4_28
13. Çiçek, Ö., Abdulkadir, A., Lienkamp, S.S., Brox, T., Ronneberger, O.: 3D U-Net: learning dense volumetric segmentation from sparse annotation. In: Ourselin, S., Joskowicz, L., Sabuncu, M.R., Unal, G., Wells, W. (eds.) MICCAI 2016. LNCS, vol. 9901, pp. 424–432. Springer, Cham (2016). https://doi.org/10.1007/978-3-319-46723-8_49
14. Milletari, F., Navab, N., Ahmadi, S.-A.: V-Net: fully convolutional neural networks for volumetric medical image segmentation. In: Proceedings of 2016 Fourth International Conference on 3D Vision, pp. 565–571. IEEE (2016)

Less is More: Contrast Attention Assisted U-Net for Kidney, Tumor and Cyst Segmentations

Mengran Wu and Zhiyang Liu[(✉)]

Tianjin Key Laboratory of Optoelectronic Sensor and Sensing Network Technology, College of Electronic Information and Optical Engineering, Nankai University, Tianjin, China
liuzhiyang@nankai.edu.cn

Abstract. As the most successful network structure in biomedical image segmentations, U-Net has presented excellent performance in many medical image segmentation tasks. We argue that the skip connections between the encoder and decoder layers pass too many redundant information, and filtered out the unnecessary information may be helpful in improving the segmentation accuracy. In this paper, we proposed a contrast attention mechanism at the skip connection, and proposed a contrast attention U-Net for KiTS21 challenge. The proposed method is able to achieve better performance over nnU-Net models with both low resolution and full resolution inputs in the 5-fold cross validation in the training set. In the final unrevealed testing set, our method achieves a mean sampled average Dice coefficient of 0.8815 and a mean sampled average surface Dice coefficient of 0.8007, which ranked the 5-th in the KiTS21 challenge.

Keywords: Kidney segmentation · Deep learning · 3D U-Net · Contrast attention

1 Introduction

Precise and quantitative evaluation on kidney masses, including tumors and cysts, have been an effective way in guiding future treatments. In clinical practice, such task is usually performed manually, which is time consuming and tedious. To make it reproducible, it is urgent to develop an automatic method for kidney mass segmentation and distinguish between tumors and cysts. Deep learning has been widely adopted in biomedical image segmentation, and has shown much higher accuracy in many tasks [1–4].

One of the most commonly adopted convolutional neural network (CNN) structure in biomedical image segmentation is U-Net [5], which have shown great performance in almost all biomedical image segmentation tasks. The skip connections between the encoder and decoder layers enable it to gather both semantic information from the deeper layers and the spatial information from the shallower layers. The nnU-Net [6], on the other hand, utilizes the U-Net structure, but focused on the preprocessing and data augmentation methods and built an self-adaptive framework for 3D medical image segmentation tasks. The success of nnU-Net not only proved the strong representation ability of U-Net, but also highlighted the importance of proper preprocessing and data augmentation methods, especially when the images are anisotropic.

© Springer Nature Switzerland AG 2022
N. Heller et al. (Eds.): KiTS 2021, LNCS 13168, pp. 46–52, 2022.
https://doi.org/10.1007/978-3-030-98385-7_7

From the released training dataset of the kidney tumor segmentation (KiTS) challenge in 2021, we made several observations. 1) The CT images are anisotropic, and the slice spacing vary from about 0.5 mm to 5 mm; 2) the physical field of view varies, where some images include lung, while some only include abdomen organs; and 3) the kidney segmentation is much easier than the kidney masses. Motivated by the above observations, we propose to adopt a two-stage segmentation method as shown in Fig. 1. In particular, we extract the coarse kidney region from down-sampled low-resolution images in the first stage, and generate the fine segmentation results in the second stage.

To better segment the foregrounds, attention mechanism was further introduced to the plain U-Net. Attention mechanism has shown to be effective in both natural language processing and computer vision. In medical image segmentation tasks, several variants of U-Net have been proposed by introducing attention mechanism to U-Net [1, 7, 8]. The attention mechanism provides a guidance to the network in focusing on the most important features, and is able to increase the segmentation accuracy. Moreover, by guiding the network to extract important features, the attention mechanism is helpful in increasing the parameter efficiency. In this paper, we proposed a contrast attention (CA) mechanism to the skip connections of U-Net. Instead of passing the entire output feature maps from the encoder layers, the proposed CA modules perform as edge detectors, and only passes the local differential information to the corresponding layers at the decoder. As we will show in this paper, the proposed CAU-Net is able to overperform nnUNet in the 5-fold cross validation on the KiTS21 training set by using much fewer parameters. In the testing set, the proposed method achieves a mean sampled average dice coefficient of 0.8815 and a mean sampled average surface dice coefficient of 0.8007, which ranked at the 5-th place in the KiTS21 challenge.

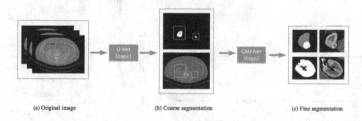

(a) Original image   (b) Coarse segmentation   (c) Fine segmentation

Fig. 1. Whole pipeline of the proposed segmentation method.

2 Methods

In this paper, we proposed a two-stage segmentation method, as shown in Fig. 1. In particular, in the first stage, coarse segmentation results for kidney were generated, which are used for extracting the regions of interest (ROIs), i.e., the regions with kidney, from the CT images. In the second stage, the CNN only focused on the ROIs and generated the fine segmentation map. In the first stage, we adopted a 3D U-Net for coarse segmentation, and in the second stage, we adopted the proposed CAU-Net for fine segmentation. The network structures of these two stages are shown in Fig. 2 and Fig. 3, respectively.

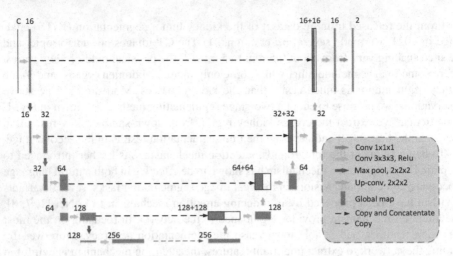

Fig. 2. 3D U-Net structure for coarse segmentation.

2.1 Training and Validation Data

Our submission made use of the official KiTS21 training set alone.

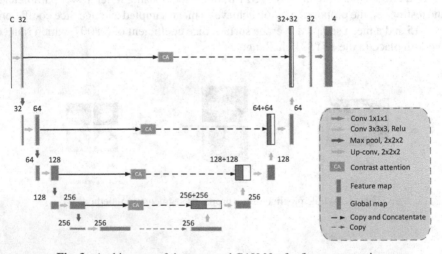

Fig. 3. Architecture of the proposed CAU-Net for fine segmentation.

2.2 Preprocessing

In KiTS21, the images varies in voxel spacing. As the CNNs are not capable in interpreting the voxel spacing, we resampled all images to a common voxel spacing. The choice of voxel spacing is in general a tradeoff between the textural information and spatial contextual information, due to the fact that the training of 3D CNNs are patch-based, instead of image-based. For instance, with a smaller voxel spacing, despite that a richer textural information can be preserved, the image patch of a certain matrix size would correspond to a smaller physical volume, leading to a small physical field of view in the CNNs. Therefore, the voxel spacing should be carefully chosen to arrive at a good tradeoff.

In particular, in the first stage, as the goal is to locate the kidney area from a coarse segmentation map, we adopted a larger voxel spacing of $3.4 \times 1.7 \times 1.7$ mm. In the second stage, to generate a fine segmentation result, a voxel spacing of $0.85 \times 0.85 \times 0.85$ mm was adopted. During resampling, the masks were resampled using nearest neighbor interpolation. The images, however, adopted varies interpolation methods on different directions. By noting that the images in the training set are generally with small voxel spacing in the transverse plane (y-z plane), bilinear interpolation was adopted on the y-z plane. In the x-axis, as the voxel spacing varies from 0.5mm to 5mm, to reduce resampling artifacts, nearest neighbor interpolation was adopted.

In CT images, the image intensity values are in fact the HU values, and the typical HU values of different tissues in CT images are presented in Table 1. In this paper, we clipped the HU values to the range $[-79, 304]$. Then the values were subtracted by 101 and divided by 76.9. All samples in the training set were used.

In our experiment, the images were cropped to patches with size (144, 128, 128) in the first stage, and size (96, 96, 96) in the second stage. In the second stage, to balance the patches with foreground and background, a dedicated patch sampler was adopted to ensure that at least one third of the patches are centered at the foreground.

Table 1. Typical tissues radiodensities of human body [1].

Tissue	HU
Air	−200
Bone	400+
Kidney	25–50
Water	0 ± 10
Blood	3–14

2.3 Proposed Network Architecture

The proposed Contrast Attention U-Net (CAU-Net) employs a U-Net like structure in general. In the classical U-Net, skip connections are employed to fuse the feature maps hierarchically with the decoder feature maps. In our proposed network, contrast attention (CA) module is added at the skip connections to encourage the network extract the most prominent features and pass them to the decoder. As shown in Fig. 3, the proposed Contrast Attention U-Net (CAU-Net) employs a U-Net like structure in general, but makes several important modifications. The detail hyperparameters, such as strides and kernel sizes can be found in Fig. 3.

2.3.1 Contrast Attention Module

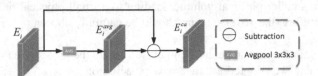

Fig. 4. The architecture of the contrast attention module.

U-Net is the most successful network architecture in medical image segmentation, which fuses high-level and low-level features by the skipping connections to obtain rich contextual information and precise location information. However, simply concatenating or uniform weighting different levels of feature maps may introduce a large amount of redundant information, which may lead to the blurring of the extracted image features, even smooth boundaries. To cope with this problem, we added the contrast attention (CA) at the skip connections of a classical U-Net, which can remove the identical information and extract the local differential information from the feature maps. The feature values of the same tissues are similar and the feature values of different tissues are quite different. Therefore the CA module is also equivalent to an implicit edge attention module, which can make the model better distinguish different tissues.

Figure 4 presented the design of the CA module. The output feature of the i-th layer can be obtained as

$$D_i = cat(up(D_{i+1}), E_i^{ca}) \tag{1}$$

for $i = 1, 2, 3, 4$, where cat denotes the channel-wise concatenation, up denotes bilinear upsampling. D_i represent the i-th decoder feature map. E_i^{ca} denotes the i-th CA feature map, and it is defined as

$$E_i^{ca} = E_i - Avg(E_i) \tag{2}$$

where E_i denotes the encoder feature map of the i-th layer. Avg represents an average pooling layer with a window size of 3 and stride 1.

2.3.2 Loss Function

The sum of dice loss and cross entropy loss is adopted as loss function. The total loss is the average of the 5 deep supervision losses.

2.3.3 Training and Validation Strategies

We adopted stochastic gradient descent (SGD) with Nestrov trick as the optimizer, with an initial learning rate of 0.01 and momentum 0.99. The batch size is set to be 2. We define an epoch as 250 batch iterations. The learning rate reduces in a polynomial way. The network is trained for 1000 epochs. Data augmentation is also implemented to reduce the over-fitting risk, with random zooming, random rotation, random flipping, and random Gaussian smoothing adopted. But we switched off the data augmentation in the last 10 epochs.

Five-fold cross validation is adopted. At each fold, we monitored the dice coefficients of the last 50 epochs, and selected the result of the last epoch as the final model.

2.3.4 Ensembling and Post-processing

Test-time augmentation is also adopted by mirroring the images along the three axes. In the first stage, we keep the largest two components, and crop the regions of interest, which is used as the input of the second stage. It is worth noting that if one of the components is much smaller than the other, only the largest component is retained. In the second stage, we use the same strategy to remove some false positives. To generate the final segmentation results in the testing set, we further use majority voting to ensemble the results of the five final models in the five-fold cross validation.

3 Results

The network is trained on a workstation with Nvidia GTX 1080Ti GPU with 11GB memory. Due to limited memory, the batch size is set to be 2. The network is implemented on PyTorch v1.9.0 [9] and monai v0.5.3 [10]. Each training epoch took about 250 s, and the training for each fold took about 84 h. During inference, the time consumed for each subject is about 60 s.

Table 2 summarizes the five-fold cross validation results on the trining set. As we can see from Table 2, the proposed method achieved better performance than full resolution and low resolution nnU-Net in all metrics. Compared with cascade nnU-Net, the method was worse in kidney and kidney mass segmentation, but better in tumor segmentation. It is worth noting that our method only has 25M parameters in total, which is far less than nnU-Net.

Table 2. Five-fold cross validation results on the KiTS21 training set.

Method	# parameters	Kidney DC	Mass DC	Tumor DC	Kidney SD	Mass SD	Tumor SD
nnUNet(full)	31.2M	0.9666	0.8618	0.8493	0.9336	0.7532	0.7371
nnUNet(low)	31.2M	0.9683	0.8702	0.8508	0.9272	0.7507	0.7347
nnUNet(cascade)	64.4M	**0.9747**	**0.8799**	0.8491	**0.9453**	**0.7714**	0.7393
Proposed	5.6M(stage1) 18.87M(stage2)	0.9693	0.8760	**0.8509**	0.9357	0.7686	**0.7449**

We tested our method on the 100 CT scans of the KiTS21 Challenge, the performance is shown in Table 3.

Table 3. The results of our model on the KiTS21 test dataset.

Method	Mean sampled average dice	Mean sampled average SD
Proposed	0.8815	0.8007

4 Discussion and Conclusion

In this paper, a CAU-Net was proposed for KiTS21. The CA modules are proposed to remove redundant information and extract the local differential information. With CA module on the skip connection, less information has been passed to the decoder layers, but the edge information is more prominent and the performance of the model is better. In the KiTS21 test dataset, the results of our method are 0.8815 for the Mean Sampled Average Dice and 0.8007 for the Mean Sampled Average Dice.

Acknowledgment. This work is supported in part by the National Natural Science Foundation of China (61871239).

References

1. Jin, Q., et al.: RA-UNet: a hybrid deep attention-aware network to extract liver and tumor in CT scans. Front. Bioeng. Biotechnol. (2020)
2. Valanarasu, J.M.J., Sindagi, V.A., Hacihaliloglu, I., Patel, V.M.: KiU-Net: towards accurate segmentation of biomedical images using over-complete representations. In: Martel, A.L., et al. (eds.) MICCAI 2020. LNCS, vol. 12264, pp. 363–373. Springer, Cham (2020). https://doi.org/10.1007/978-3-030-59719-1_36
3. Li, W.J., Jia, F., Hu, Q.: Automatic segmentation of liver tumor in CT images with deep convolutional neural networks. J. Comput. Chem. **03**, 146–151 (2015)
4. Thaha, M.M., Kumar, K.P.M., Murugan, B.S., Dhanasekeran, S., Vijayakarthick, P., Selvi, A.S.: Brain tumor segmentation using convolutional neural networks in MRI images. J. Med. Syst. **43**(9), 1 (2019). https://doi.org/10.1007/s10916-019-1416-0
5. Ronneberger, O., Fischer, P., Brox, T.: U-Net: convolutional networks for biomedical image segmentation. In: Navab, N., Hornegger, J., Wells, W.M., Frangi, A.F. (eds.) MICCAI 2015. LNCS, vol. 9351, pp. 234–241. Springer, Cham (2015). https://doi.org/10.1007/978-3-319-24574-4_28
6. Isensee, F., et al.: nnU-Net: Breaking the Spell on Successful Medical Image Segmentation. arXiv preprint arXiv:1904.08128 (2019)
7. Oktay, O., et al.: Attention U-Net: Learning Where to Look for the Pancreas. arXiv preprint arXiv:1804.03999 (2018)
8. Li, C., et al.: Attention Unet++: a nested attention-aware U-Net for liver CT image segmentation. In: 2020 IEEE International Conference on Image Processing (ICIP), pp. 345–349 (2020)
9. Paszke, A., et al.: Automatic differentiation in PyTorch (2017)
10. Project MONAI Medical Open Network for AI. https://docs.monai.io/en/latest/index.html

A Coarse-to-Fine Framework for the 2021 Kidney and Kidney Tumor Segmentation Challenge

Zhongchen Zhao, Huai Chen, and Lisheng Wang(✉)

Institute of Image Processing and Pattern Recognition, Department of Automation, Shanghai Jiao Tong University, Shanghai 200240, People's Republic of China
{13193491346,lswang}@sjtu.edu.cn

Abstract. Kidney cancer is one of the most common malignant tumors in the world. Automatic segmentation of kidney, kidney tumor, and kidney cyst is a essential tool for kidney cancer surgery. In this paper, we use a coarse-to-fine framework which is based on the nnU-Net and achieve accurate and fast segmentation of the kidney and kidney mass. The average Dice and surface Dice of segmentation predicted by our method on the test are 0.9077 and 0.8262, respectively. Our method outperformed all other teams and achieved 1st in the KITS2021 challenge.

Keywords: Automatic kidney segmentation · Kidney cancer · Coarse-to-fine framework

1 Introduction

Kidney cancer is the 13th most common cancer worldwide, accounting for 2.4% of all cancers, with more than 330,000 new cases diagnosed yearly, and its incidence is still increasing [1]. Due to the wide variety in kidney tumor morphology, it's laborious work for radiologists and surgeons to delineate the kidney and its mass manually. Besides, the work relies on assessments that are often subjective and imprecise.

Automatic segmentation of renal tumors and surrounding anatomy is a promising tool for addressing these limitations: Segmentation-based assessments are objective and necessarily well-defined, and automation eliminates all effort save for the click of a button. Expanding on the 2019 Kidney Tumor Segmentation Challenge [2], KiTS2021 aims to accelerate the development of reliable tools to address this need, while also serving as a high-quality benchmark for competing approaches to segmentation methods generally.

Supplementary Information The online version contains supplementary material available at https://doi.org/10.1007/978-3-030-98385-7_8.

2   Methods

Semantic segmentation of organs is one of the most common tasks in medical image analysis. There are already many accurate and efficient algorithms for medical image segmentation tasks, such as nnU-Net [3]. In this paper, we use the nnU-Net as a baseline and adopt the coarse-to-fine strategy to segment the kidney, the kidney tumor, and the kidney cyst, as shown Fig. 1. We also propose a surface loss to make the network segment the surface better. To be specific, our algorithm contains three steps:

Coarse Segmentation. We first use a nnU-net to get the coarse segmentation. Then we crop the tightest bounding box (bbox) containing kidney region-of-interest and expand the bbox 1.5 times. The kidney mass is always contained in the kidney area. So we can use the kidney ROI, instead of the original full CT image, to get more accurate segmentation results. This stepis is necessary to crop image to a smaller size and retain useful areas.

Fine Kidney Segmentation. Then we refine the predictions of kidney from the cropped kidney ROI by a single classification nnU-net.

Fine Tumor and Mass Segmentation. With the kidney ROI and refine kidney segmentation, we segment the kidney tumor and mass by two nnU-Net separately and combine them with the refine kidney segmentation as the final segmentation.

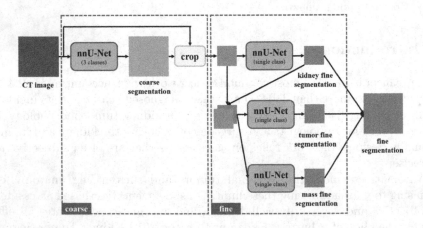

Fig. 1. An overview of our coarse-to-fine segmentation framework.

2.1   Training and Validation Data

Our submission use the official KiTS2021 training set alone [4]. We divide the provided data into training set and validation set at a ratio of 4:1.

2.2   Preprocessing

In each case, the labels are annotated by three independent people and we use the majority voting strategy for the multiple annotations. We follow the way in the nnU-Net to preprocess the training data. The spacing of all official CT images is the same on the x-axis and y-axis, but different on the z-axis. We resample all images to the same target spacing [0.786 0.78125 0.78125] using third order spline interpolation and then normalize the data. Besides, A variety of data augmentation techniques are applied on the fly during training: rotations, scaling, mirroring, etc.

2.3   Proposed Method

2.3.1   Kidney Segmentation

After hard mining, we find the accuracy of mass is worse than the accuracy of the kidney. So we decide to segment the kidney mask firstly and then segment the tumor and cyst from the kidney mask. Specifically, we first use the nnU-Net to get the kidney ROI from the CT image for each case and then use the coarse-to-fine framework to segment the kidney mask from the kidney ROI.

2.3.2   Mass Segmentation

With the predicted kidney mask, we use another nnU-Net to segment the mass. By analyzing the provided CT images, we find that the mass is more likely to be on the edge of the kidney. Therefore, we combine the predicted kidney mask, which contains the kidney edge information, with the kidney ROI and feed them to the nnU-Net.

2.3.3   Tumor Segmentation

To further improve the tumor segmentation performance, we train another nnU-Net to segment the tumor alone. Similar to the mass Segmentation, we also use the predicted kidney mask and the kidney ROI as the inputs for the nnU-Net.

2.3.4   Surface Loss

Compared to the 2019 KiTS Challenge, the metric in this year has one more Surface Dice [5], which is used to quantitatively assess the overlap between the predicted surface of segmentation and the real surface. Considering that, we propose the surface loss function, which will penalize the unacceptable regions of the predicted surface. The surface loss function is defined as the summation of distances from all false-positive points and false-negative points in the prediction to the surface of ground truth, as shown below:

$$L_s = \frac{1}{C} \sum_{p_{pred} \in FP \cup FN} \left(\min_{p_{gt} \in S_{gt}} \|p_{pred} - p_{gt}\|_2 \right) \tag{1}$$

where S_{gt} is the surface of the ground truth, FP and FN are the sets of false-positive points and false-negative points separately, and C is a constant.

During the actual training process, we combine the surface loss with the Dice loss and cross-entropy loss, and do not use the surface loss until the Dice loss and cross-entropy loss are low enough. In other words, the surface loss is used to fine-tune the final surface segmentation results in our method.

2.3.5   Postprocessing

After training, we remove the isolated blocks which are smaller than 20,000 voxels for the kidney. We also remove all tumors and cysts outside kidneys and take the rest as the final segmentation results. Besides, we combine the kidney coarse segmentation and kidney fine segmentation by voting, as the final kidney prediction. We also combine the tumor fine segmentation and the tumor segmentation from the mass fine segmentation to get better tumor prediction.

2.3.6   Implementation Details

During the training, we use batch normalization (BN) where the batch size is 2. And we use stochastic gradient descent (SGD) as the optimizer. The initial learning rate is set to be 0.01 and the total training epochs are 1000. The patch size in the coarse segmentation stage is [128,128,128] while [128,192,96] in the fine segmentation stage. Considering the training time consuming, we do not use the 5-fold cross-validation as mentioned in nnU-Net [3], and just split all training data into training set and validation set randomly. Other hyper-parameters mainly follow the nnU-Net as default. We implement our network with PyTorch based on a single NVIDIA GeForce RTX 3090 GPU with 24 GB memory.

3   Results

3.1   Metric

According to the organizers' request, we use two metrics for evaluation, the volumetric Dice coefficient, and the Surface Dice. And we use Hierarchical Evaluation Classes (HECs) for the three classes of targets, in which classes that are considered subsets of another class are combined with that class for the purposes of computing a metric for the superset. For KiTS2021, the following HECs will be used:

Kidney and Masses: Kidney + Tumor + Cyst
Kidney Mass: Tumor + Cyst
Tumor: Tumor only

3.2   Results and Discussions

We summarize the quantitative results on in Table 1. All the results are based on the validation set, which contains 60 cases. The average Dice is 0.9099, and the Dice for kidney, kidney mass, kidney tumor are 0.9752, 0.8851, and 0.8693 respectively. The average Surface Dice is 0.8348, and the Surface Dice for kidney, kidney mass, kidney tumor are 0.9486, 0.7867, and 0.7692 respectively. For the tumor segmentation, our algorithm performs significantly better than the baseline. While for the kidney and cyst segmentation, the improvement of our algorithm is very small. This is because we didn't use the cascaded nnU-Net due to its long training time.

On the test set, the average dice of our prediction is 0.9077 and the average surface dice is 0.8262, as shown in Table 2. Compared to the 2nd method, our method segment the kidney tumor more accurately by 2.8%, resulting in a nearly 1% improvement in both Dice and Surface Dice. We believe this is mainly because we use the kidney fine prediction as an additional input for tumor segmentation to provide the kidney information.

Figure 2 shows a demonstration of the prediction result. The left side is the ground truth and the right side is our prediction.

Table 1. Results of experiments on validation set. (Splitted by ourself)

Model	Dice				Surface dice			
	kidney	Mass	Tumor	Ave	Kidney	Mass	Tumor	Ave
Baseline	0.9748	0.8793	0.8365	0.8969	0.9499	0.7810	0.7392	0.8234
Ours	0.9752	0.8851	0.8693	**0.9099**	0.9486	0.7867	0.7692	**0.8348**
Improvement	+0.04%	+0.58%	+3.28%	**+1.30%**	−0.13%	+0.57%	+3.00%	**+1.14%**

Table 2. Quantitative results on test set. (Given by the organizer)

Team	Rank	Dice	Surface dice	Tumor dice
Ours	1	**0.908**	**0.826**	**0.860**
Alex Golts et al.	2	0.896	0.816	0.832
Yasmeen George	3	0.894	0.814	0.831

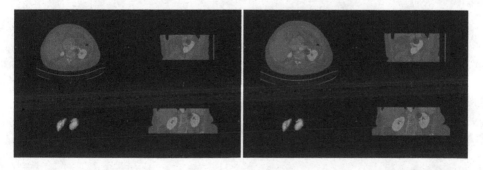

Fig. 2. A demonstration of prediction results of case 010. The kidney is shown in red and the tumor is shown in green. (Color figure online)

4 Conclusion

In this paper, we use a coarse-to-fine framework to segment the kidney, tumor, and cyst from CT images. We use the nnU-Net as a baseline and improve it by using the surface loss and ROI cropping. Experiments show that our method truly works. The Dice of the kidney is very high and we segment the kidney tumor more accurately. Our method outperforms all other competing teams, and we attribute this mainly to the three parts: the good performance of the nnU-Net; using kidney segmentation to predict tumor and mass; our postprocess. Finally, we hope that our work can make some contribution to the automatic segmentation for kidney cancer.

Acknowledgment. We would like to express our gratitude to the KiTS2021 organizers and the nnU-Net team. We also want to say thanks to Nicholas Heller and Fabian Isensee for their kind help.

References

1. Scelo, G., Larose, T.L.: Epidemiology and risk factors for kidney cancer. J. Clin. Oncol. **36**(36), 3574 (2018)
2. Heller, N., et al. The state of the art in kidney and kidney tumor segmentation in contrast-enhanced CT imaging: results of the KITS19 challenge. Med. Image Anal. **67**, 101821 (2021)
3. Isensee, F., Jaeger, P.F., Kohl, S.A.A., Petersen, J., Maier-Hein, K.H.: nnU-Net: a self-configuring method for deep learning-based biomedical image segmentation. Nat. Methods **18**(2), 203–211 (2021)
4. Heller, N., et al.: The KITS19 challenge data: 300 kidney tumor cases with clinical context, CT semantic segmentations, and surgical outcomes. arXiv preprint arXiv:1904.00445 (2019)
5. Nikolov, S., et al.: Deep learning to achieve clinically applicable segmentation of head and neck anatomy for radiotherapy. arXiv preprint arXiv:1809.04430 (2018)

Kidney and Kidney Tumor Segmentation Using a Two-Stage Cascade Framework

Chaonan Lin[1], Rongda Fu[2], and Shaohua Zheng[1(⊠)]

[1] College of Physics and Information Engineering, Fuzhou University, Fuzhou, China
sunphen@fzu.edu.cn
[2] School of Mechanical Engineering and Automation,
Fuzhou University, Fuzhou, China

Abstract. Automatic segmentation of kidney tumors and lesions in medical images is an essential measure for clinical treatment and diagnosis. In this work, we proposed a two-stage cascade network to segment three hierarchical regions: kidney, kidney tumor and cyst from CT scans. The cascade is designed to decompose the four-class segmentation problem into two segmentation subtasks. The kidney is obtained in the first stage using a modified 3D U-Net called Kidney-Net. In the second stage, we designed a fine segmentation model, which named Masses-Net to segment kidney tumor and cyst based on the kidney which obtained in the first stage. A multi-dimension feature (MDF) module is utilized to learn more spatial and contextual information. The convolutional block attention module (CBAM) also introduced to focus on the important feature. Moreover, we adopted a deep supervision mechanism for regularizing segmentation accuracy and feature learning in the decoding part. Experiments with KiTS2021 testset show that our proposed method achieve Dice, Surface Dice and Tumor Dice of 0.650, 0.518 and 0.478, respectively.

Keywords: Cascade framework · Kidney/tumor segmentation · Deep learning

1 Introduction

Kidney cancer is one of the most aggressive cancer, which have 40 0000 growth numbers and high fatality rate of 40%. Renal cell carcinoma (shorted as kidney tumor) and renal cysts (cysts for short) are the most common diseases that cause to it. The cysts are formed in the kidneys with age, and won't easily bring about injury while tumors often pose high risks to human health.

Studies have shown that tumors are more susceptible to effective treatment if they are detected at earlier stage. However, these tumors may grow into a large size before being detected [1]. Therefore, early accurate diagnosis can effectively improve the survival rate of kidney cancer patients. The success of such studies

N. Heller et al. (Eds.): KiTS 2021, LNCS 13168, pp. 59–70, 2022.
https://doi.org/10.1007/978-3-030-98385-7_9

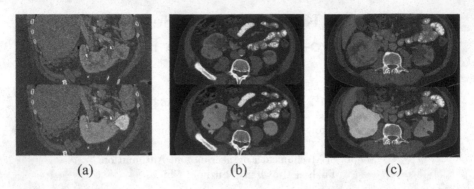

(a)   (b)   (c)

Fig. 1. Illustration of sample segmented images from three patients in the KiTS21 dataset. The first row is the transverse plane, and the second row is with the labels. Red, green and blue denote kidney, kidney tumor and cyst, respectively. (Color figure online)

relies on the computed tomography (CT) technology, which can provide high-resolution images with good anatomical details. Combined with the great potential information from medical images, such as the location, shape and size of the kidney and tumor, radiologists can know about disease severity and progression, then make a more accurate clinical decision. Hence, accurate segmentation of kidney and kidney tumor is an essential step for radiomic analysis as well as developing advanced surgical planning techniques. The segmentation of kidney, kidney tumor and cyst are usually manually marked by radiologists. However, manual segmentation is a time-consuming and tedious task due to the hundreds slices of CT. Moreover, the label results strongly depend on the experience of radiologists and prone to errors. In order to reduce the burden of manual works and improve segmentation accuracy, automatic segmentation of kidney, kidney tumor and cyst has become a new demand.

Deep learning (DL) technology has been widely applied in the medical image field and play an essential role in kidney and kidney tumor segmentation works. Methods based on deep learning are categorized into two: one-stage and two-stage method. One-stage methods [2–7] are designed to predict the multi-class results directly from whole images. Guo et al. [5] proposed an end-to-end model based on residual and attention module. Residual connection was added to each convolutional layer to generate clearer semantic features. Skip connection was also used in attention module to make the decoder focus on the segmentation target. Zhao et al. [6] presented a multi-scale supervised 3D U-Net to segment kidneys and kidney tumors from CT images. Multi-scale supervision was adopt to obtain more accurate predictions from deep layers. Sabarinathan et al. [7] presented a novel kidney tumor segmentation method. This work introduced supervision layers into the decoder part, and coordinate convolutional layer was utilized to improvise the generalization capacity of the model.

Two-stage methods [8–13] aim to solve the imbalance problem between foreground and back-ground. Those methods firstly detect the volume of interest

(VOIs), then segment the target organs from the VOIs. A typical two-stage method was proposed by Cheng et al. [8], they employed a double cascaded framework, which decomposed the complex task of multi-class segmentation into two simpler binary segmentation tasks. In the first step, the region of interest (ROI) including kidney and kidney tumor is extracted, and then segment the kidney tumor in second step. However, these works still suffer from several anatomical challenges. First, the low contrast between kidney and nearby organs, the unclear boundaries and heterogeneity of tumor, all make accurate segmentation become more difficult, as shown in Fig. 1(a) and Fig. 1(b). Second, Fig. 1(c) indicates that kidney tumors and cysts exhibit various size, shape, location and number from different patients.

To address the above challenges and improve the segmentation performance of unbalanced kidney and tumor datasets, we proposed a two-stage framework to obtain kidney and masses (kidney, cyst), respectively. In this framework, the complex multi-class segmentation task is transformed into two simplified subtasks: (i) locating the kidney region and segmenting the kidney, (ii) segmenting the kidney tumor and cyst in the kidney. The Kidney-Net applied in the first stage is modified based on a normal 3D U-Net, while our core works mainly focus on masses segmentation in the second stage. In the second stage, a fine segmentation network Masses-Net is trained based on the cropped kidney region obtained in the first stage to segment tumors and cysts. In order to leverage more useful features, a multi-dimension feature (MDF) module is utilized to learn more space and context information. Meanwhile, convolutional block attention module (CBAM) also applied to focus on the important feature. Finally, a deep supervision mechanism was also used in the decoding, which works as a regularizing role in segmentation accuracy and feature learning.

2   Methods

In this section, we mainly introduce our method for kidney and kidney masses (tumor and cyst) segmentation. The proposed two-stage segmentation framework is illustrated in Fig. 2. The framework consists of two phases: the first stage for kidney segmentation and the second stage for masses (tumor and cyst) segmentation.

In the first stage, the pre-processed CT images are fed into a kidney segmentation network, which named Kidney-Net. The output of Kidney-Net is a coarse segmentation result of overall kidney and is binarized to produce an overall kidney mask. The mask is applied for boundary coordinates and crop volume of interest (VOI). The cropped VOI is the input of the second stage, a Masses-Net is used for tumor segmentation and cyst segmentation.

In the training processing, the two networks are trained individually due to the different input patches. In the testing processing, two individual results of Kidney-Net and Masses-Net are fused via a union method, which add the two prediction results directly. Then the merge result is refined by a post-processing method.

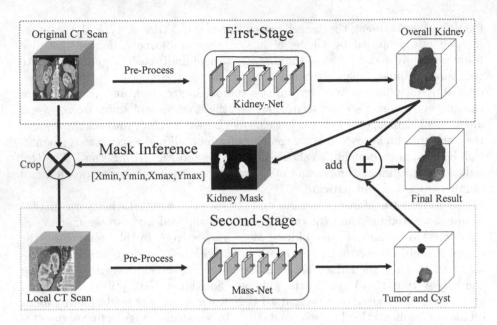

Fig. 2. Schematic of the workflow of the proposed two-stage segmentation framework.

2.1 Kidney-Net

In our method, we concatenated two networks to segment the kidney, kidney masses (tumor and cyst) respectively. The Kidney-Net in the first stage is used for kidney segmentation. As shown in Fig. 3(a), Kidney-Net is a u-shaped network which only contains three pooling layers. And the encoder/decoder blocks are all regular convolution block, which is composed of two convolution layers with batch normalization (BN) and rectified linear unit (ReLU). Due to the Kidney-Net is trained to predict the probabilities of every voxels belong to kidney, Dice loss function is utilized for the voxel level classification task and be formulated as:

$$\mathcal{L}_{kidney} = 1 - \frac{2 * \sum_{i=1}^{2} (r_i * t_i)}{\sum_{i=1}^{2} (r_i + t_i) + \theta} \tag{1}$$

Where r_i, t_i, $i \in 0, 1$ is the segmentation kidney result and target kidney mask, respectively, and θ is a smooth term to avoid division by zero.

2.2 Masses-Net

The architecture of Masses-Net is shown in Fig. 3(b). Masses-Net is similar to the Kidney-Net. For the purpose of utilizing more global features, the encoder of Masses-Net contains four pooling operations and four encoder blocks. The input and output blocks are regular convolution block, which used for generate low-level feature maps. Then, the feature maps generate by input block are fed into

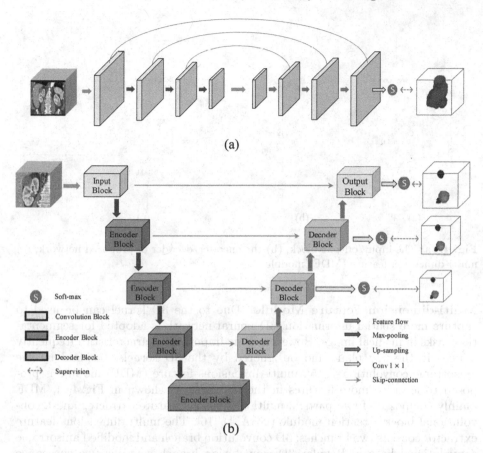

Fig. 3. Overview of the proposed segmentation network. (a) The basic architecture of the Kidney-Net (b) The basic architecture of the Masses-Net

four successive encoder blocks to obtain global features. The decoder is used for target segmentation and working in a coarse-to-fine pattern, which also contains four decoders. Finally, the output of the last three decoder blocks is fed into Soft-max activation function for the tumor and cyst prediction. Moreover, deep supervision scheme is applied. The key components are illustrated as follow.

Residual Connection Mechanism. Considering the problems related to overfitting and vanishing gradient, residual connection is incorporated to maintain more spatial and contextual information and make the learnable network parameters increasingly effective [14]. Figure 4(a) shows the input/output block, which is similar to a regular residual block. Combining the residual connection, the network can achieve a better result. Figure 4(b) shows a encoder/decoder block of our network. In this block, a multi-dimension feature (MDF) module is applied to replace the regular convolution layers. And the architecture of MDF is shown in the Fig. 4(c), residual connection also adopted in the module.

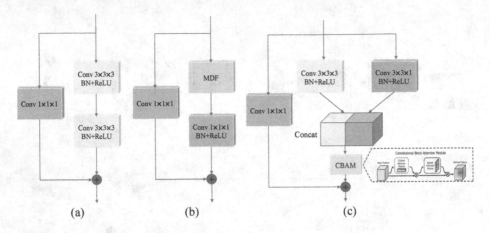

Fig. 4. (a) The input/output block, (b) the encoder/decoder of proposed network, (c) multi-dimension feature (MDF) module.

Multi-dimension Feature Module. Due to the 3D kernel can be used to capture more spatial information, 3D neural network is adopted for segmentation tasks in medical image. However, some important characteristics frequently appear in the x/y plane and be ignored by the 3D kernels. Inspired by the anisotropic convolutional [15], multi-dimensions feature (MDF) module is proposed to leverage more features in the network. As shown in Fig. 4(c), MDF mainly composed of two parts: a multi dimensions feature extractor and a convolutional block attention module (CBAM) [16]. The multi dimensions feature extractor consists two branches: 3D convolution branch and modified anisotropic convolutional branch. Regular 3D convolution branch contains two successive convolution layers with kernel of $3 \times 3 \times 3$, and is used for capturing spatial information. The modified anisotropic convolutional branch only uses the $3 \times 3 \times 1$ kernel to learn the shape characteristics in the x/y plane. In fact, the modified anisotropic convolutional operations can leverage intra-slice information without increasing the amounts of parameters too much in a 3D network. Finally, the output of two branches is concatenated, which can be formulated as:

$$m_c = Concat\left(Conv\left(3 \times 3 \times k_i\right)(x)\right) \quad i = 1, 3 \tag{2}$$

The CBAM is a lightweight attention module that combine the spatial attention and channel attention, and it can be embedded in other networks easily. CBAM contains two separate submodules: Channel Attention Module (CAM) and Spatial Attention Module (SAM), and the two submodules work respectively. The CBAM incorporated after the multi dimensions feature extractor is used to focus on the important feature in different channels from the concatenated feature maps. And the SAM is also applied to learn the region of interest which contain potential lesion characteristics. Moreover, the residual connection also applied in the MDF to leverage more feature. Finally, the output of MDF operation is formulated as:

$$Output = \mathbf{F}_{sam}\left(\mathbf{F}_{cam}\left(\mathbf{m}_c\right)\right) + x \tag{3}$$

Deep Supervision Mechanism. Increasing the depth of the neural network can improve the representation ability, while cause vanishing gradient problem to make the network become difficult to train. Deep supervision is proposed to mitigate such problems. Unlike the previous networks, deep supervision provides integrated direct supervision to the hidden layer instead of only providing supervision in the output layer and passing it back to the earlier layer. This can effectively solve the problems of gradient disappearance and slow convergence. As shown in Fig. 3, we introduced deep supervision in the last three decoder blocks of the decoder part. The output layer and the hidden layer can be supervised via deep supervision at the same time, which can jointly improve the gradient propagation to minimize the loss.

2.3 Loss Function

In order to alleviate the imbalance problem from multi-class segmentation task, the combination of dice loss and weight cross entropy (WCE) loss is utilized in our network. The loss function can be formulated by the following:

$$\mathcal{L}_{mass} = \alpha\mathcal{L}_{dice} + (1-\alpha)\mathcal{L}\mathcal{L}_{wce} \tag{4}$$

where α is the weight and be set to 0.5 in our experiment. And the formulation of L_{dice} and L_{wce} can be describe detail as follow:

$$\mathcal{L}_{dice} = 1 - \frac{2 * \sum_{i=1}^{3}\sum_{n=1}^{N}\left(r_{i_n} * t_{i_n}\right)}{3 * \sum_{i=1}^{3}\sum_{n=1}^{N}\left(r_{i_n} + t_{i_n}\right) + \theta} \tag{5}$$

$$\mathcal{L}_{wce} = \sum_{i=0}^{3} \mathrm{w}_i \sum_{n=1}^{N}\left(r_{i_n}\log\left(t_{i_n}\right) + \left((1-r_{i_n})\log\left(1-t_{i_n}\right)\right)\right) \tag{6}$$

Where N denote the voxel number, and i denote the index of each voxel. r_{i_n} and t_{i_n} is the predicted result and the target label of voxel n on category i. And w_i is the weight in the weight cross entropy (WCE) loss.

As shown in Fig. 4, the deep supervision is adopted for the multi outputs in the networks. The outputs from various scales are up-sampled to the original image size. Hence, the final loss in out training stage is formulated as:

$$\mathcal{L}_{total} = \frac{1}{3}\sum_{l=2}^{l=4}\mathcal{L}_{mass}\left(R_l, T\right) \tag{7}$$

3   Experiment

3.1   Datasets

The abdominal CT of 300 patients from KiTS21 challenge are applied for training and evaluation in our experiment. Kidney Tumor Segmentation (KiTS) dataset was collected from either an M Health Fairview or Cleveland Clinic medical center between 2010 and 2020. The dataset provides the segmentation labels which contain four classes: (i) background, (ii) kidney, (iii) tumor and (iv) cyst. And the ground-truth of each CT are annotated by professional medical experts. It should be noted that excluding and modifying training cases was explicitly permitted. Therefore, we excluded 10 cases, which some slices are contaminated and loss some information so that unable to load and train by our network. The IDs of these cases are $52, 60, 65, 66, 91, 111, 115, 135, 140, 150$. Finally, the rest 290 CT scans were randomly split into training set and validation set with a radio of 4:1. Due to use cross-validation would force the computation time to be multiplied by the number of folds, we only keep a validation set rather than using cross-validation. In order to strike a balance between training enough data and being able to predict our errors, 20% is selected as the account of validation. In the experiments, we train the model and fine-tune hyper-parameter on the training set, and the validation set was used for chose the model with best results.

3.2   Pre-processing and Post-processing

One challenge in kidney and tumor segmentation is the unclear boundary. The mitigate the problem, Gaussian filter is applied, then CT voxel intensity of the filtered images is clipped into $[-100, 400]$. Finally, the images are normalized via the z-score normalization. Considering the limit of GPU memory, a sliding window technique is adopted to crop the whole CT image into smaller patches. In the first stage, the size of patches is $256 \times 256 \times 16$ and the stride is $[128, 128, 8]$. For the second stage, we cropped patches in the VOI after mask inference. And the size of final training patches is $96 \times 96 \times 64$ with a stride of $[48, 48, 32]$.

In the post-processing, the connected component analysis is utilized to remove the unconfident candidates and keep the largest two components as left and right kidney.

3.3   Training and Implementation Details

The proposed networks were implemented using Python based on the Pytorch and experiments were performed on a computer with a single GPU (i.e., NVIDIA GTX 1080 Ti) and Linux Ubuntu 18.04 LTS 64-bit operating system. We use Adam optimizer ($\beta 1 = 0.9$, $\beta 2 = 0.999$) with initial learning rate 1.0–04 for optimization in the training stage. And the batch-size and training epochs are set as 2 and 50 respectively.

3.4   Metrics

Our method was evaluated by its Sørensen-Dice score and surface Dice. The Sørensen-Dice are defined by the following formulas:

$$D(P,G) = \frac{2 \times |P \cap G|}{|P| + |G|} \tag{8}$$

where P and G represent the predicted segmentation results and the ground truth, respectively. It should be noted that the quantitative results in our experiment are calculated by the evaluation codes, which provided by the challenge organizations.

4   Results and Discussion

Table 1 shows the results of the ablation study for our proposed method and highlights the effect of each component applied to the model on the segmentation results. We evaluate the performance of each component by removing the multi-dimension feature module (MDF) and deep supervision (DS), respectively. We exploit Sørensen-Dice and surface Dice as the evaluation metric and finally report the average of all cases.

From the table it is observed that, during validation, the proposed method achieves the Sørensen-Dice score of 0.9304, 0.5729 and 0.563 for kidney, masses and kidney tumor respectively by involving multi-dimension feature module (MDF) and deep supervision mechanism. Similarly, our network without MDF got separately Sørensen-Dice score of 0.9321, 0.5535 and 0.5463 for the three-class regions. It can also be found that, without incorporating DS in the proposed architecture, Sørensen-Dice score is reduced to 0.9306, 0.5519 and 0.5262 respectively for kidney, masses and kidney tumor.

In the testset of KiTS21, we achieve the Dice, Surface Dice and Tunor Dice of 0.650, 0.518 and 0.478, respectively.

The qualitative results of KiTS21 dataset on our proposed model is shown in Fig. 5. The first column shows the ground truth of input images. The rest columns show the results of our ablation study. In the output images, the green and blue colored spot are tumor region and cyst region, whereas the red color spot is the kidney region. From the qualitative results it is observed that the efficacy of our proposed network.

Table 1. Ablation study on the KiTS validation dataset for the multi-dimension feature module (MDF) and deep supervision (DS) into the baseline framework.

Method	Dice kidney	Dice masses	Dice tumor	SD kidney	SD masses	SD tumor
Ours	0.9304	0.5729	0.563	0.8722	0.4006	0.3987
Ours w/o MDF	0.9321	0.5535	0.5463	0.8758	0.3765	0.3784
Ours w/o DS	0.9306	0.5519	0.5262	0.8723	0.3783	0.3687
U-Net	0.9307	0.5333	0.5209	0.8722	0.3658	0.364

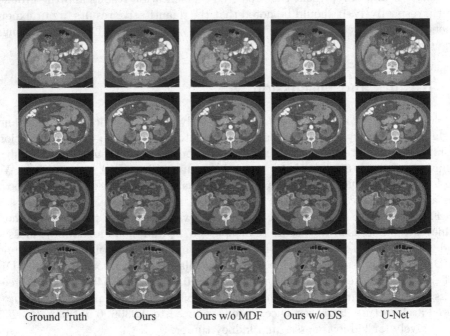

Ground Truth Ours Ours w/o MDF Ours w/o DS U-Net

Fig. 5. Examples of segmentation results. Each row denotes one patient, and from left to right, each column represents the ground truth and different predictions by our method, our network without MDF, our network without DS and U-Net, respectively. The red mask and green mask represent separately kidney area and masses (tumor and cyst) area, while the blue mask represents cyst area. (Color figure online)

5   Conclusion

In this paper, we described a two-stage cascade framework to obtain kidney, kidney tumor and cyst, respectively. The complex multi-class segmentation task is translated into two subtasks. We designed two networks to implement the subtasks and named them Kidney-Net and Masses-Net, respectively. Kidney-Net is used for kidney segmentation while Masses-Net is applied for masses (tumor and cyst) segmentation. We merge the outputs from two networks and the merge result is refined via a post-processing method. In order to leverage more space and context information, a multi-dimension feature (MDF) module is embedded into the Mass-Net. And convolutional block attention module (CBAM) also applied to learn important feature. In addition, deep supervision mechanism is utilized for improving segmentation accuracy.

Our method segment three critical organs on KiTS21 validation dataset with the Sørensen-Dice of 0.9304, 0.5729 and 0.563, respectively. While our method has shown effectiveness on the validation set, we will continue to work on further optimization of the network in the future.

Acknowledgment. This work was supported by the Fujian Provincial Natural Science Foundation project (2021J01578, 2021J02019, 2019Y9070).

References

1. Choudhari, K., Sharma, R., Halarnkar, P.: Kidney and tumor segmentation using U-Net deep learning model. In: 5th International Conference on Next Generation Computing Technologies (NGCT 2019) (2020)
2. Yang, G., Li, G., Pan, T., Kong, Y., Zhu, X.: Automatic segmentation of kidney and renal tumor in CT images based on 3D fully convolutional neural network with pyramid pooling module. In: 2018 24th International Conference on Pattern Recognition (ICPR) (2018)
3. Yu, Q., Shi, Y., Sun, J., Gao, Y., Zhu, J., Dai, Y.: Crossbar-Net: a novel convolutional neural network for kidney tumor segmentation in CT images. IEEE Trans. Image Process. **28**, 4060–4074 (2019)
4. Isensee, F., Maier-Hein, K.H.: An attempt at beating the 3D U-Net (2019)
5. Guo, J., Zeng, W., Yu, S., Xiao, J.: RAU-Net: U-Net model based on residual and attention for kidney and kidney tumor segmentation. In: 2021 IEEE International Conference on Consumer Electronics and Computer Engineering (ICCECE), pp. 353–356. IEEE (2021)
6. Zhao, W., Jiang, D., Queralta, J.P., Westerlund, T.: MSS U-Net: 3D segmentation of kidneys and tumors from CT images with a multi-scale supervised U-Net. Inform. Med. Unlocked **19**, 100357 (2020)
7. Sabarinathan, D., Parisa Beham, M., Mansoor Roomi, S.M.M.: Hyper vision net: kidney tumor segmentation using coordinate convolutional layer and attention unit. In: Babu, R.V., Prasanna, M., Namboodiri, V.P. (eds.) NCVPRIPG 2019. CCIS, vol. 1249, pp. 609–618. Springer, Singapore (2020). https://doi.org/10.1007/978-981-15-8697-2_57
8. Cheng, J., Liu, J., et al.: A double cascaded framework based on 3D SEAU-Net for kidney and kidney tumor segmentation (2019)
9. Hou, X., Xie, C., Li, F., Wang, J., Nan, Y.: A triple-stage self-guided network for kidney tumor segmentation. In: 2020 IEEE 17th International Symposium on Biomedical Imaging (ISBI) (2020)
10. Causey, J., et al.: An ensemble of u-net models for kidney tumor segmentation with CT images. IEEE/ACM Trans. Comput. Biol. Bioinform. (2021)
11. Zhang, Y., Wang, Y., Hou, F., et al.: Cascaded volumetric convolutional network for kidney tumor segmentation from CT volumes (2019)
12. Xie, X., Li, L., Lian, S., Chen, S., Luo, Z.: SERU: a cascaded SE-ResNeXT U-Net for kidney and tumor segmentation. Concurr. Comput. Pract. Exp. **32**(2), 5738 (2020)

13. Yan, X., Yuan, K., Zhao, W., Wang, S., Cui, S.: An efficient hybrid model for kidney tumor segmentation in CT images. In: 2020 IEEE 17th International Symposium on Biomedical Imaging (ISBI) (2020)
14. He, K., Zhang, X., Ren, S., Sun, J.: Deep residual learning for image recognition. In: Proceedings of the IEEE Conference on Computer Vision and Pattern Recognition, pp. 770–778 (2016)
15. Wang, G., Li, W., Ourselin, S., Vercauteren, T.: Automatic brain tumor segmentation using cascaded anisotropic convolutional neural networks. In: Crimi, A., Bakas, S., Kuijf, H., Menze, B., Reyes, M. (eds.) BrainLes 2017. LNCS, vol. 10670, pp. 178–190. Springer, Cham (2018). https://doi.org/10.1007/978-3-319-75238-9_16
16. Woo, S., Park, J., Lee, J.-Y., Kweon, I.S.: CBAM: convolutional block attention module. In: Ferrari, V., Hebert, M., Sminchisescu, C., Weiss, Y. (eds.) ECCV 2018. LNCS, vol. 11211, pp. 3–19. Springer, Cham (2018). https://doi.org/10.1007/978-3-030-01234-2_1

Squeeze-and-Excitation Encoder-Decoder Network for Kidney and Kidney Tumor Segmentation in CT Images

Jianhui Wen, Zhaopei Li, Zhiqiang Shen, Yaoyong Zheng,
and Shaohua Zheng$^{(\boxtimes)}$

College of Physics and Information Engineering, Fuzhou University, Fuzhou, China
sunphen@fzu.edu.cn

Abstract. Kidney cancer is one of the top ten cancers in the world, and its incidence is still increasing. Early detection and accurate treatment are the most effective control methods. The precise and automatic segmentation of kidney tumors in computed tomography (CT) is an important prerequisite for medical methods such as pathological localization and radiotherapy planning, However, due to the large differences in the shape, size, and location of kidney tumors, the accurate and automatic segmentation of kidney tumors still encounter great challenges. Recently, U-Net and its variants have been adopted to solve medical image segmentation problems. Although these methods achieved favorable performance, the long-range dependencies of feature maps learned by convolutional neural network (CNN) are overlooked, which leaves room for further improvement. In this paper, we propose an squeeze-and-excitation encoder-decoder network, named SeResUNet, for kidney and kidney tumor segmentation. SeResUNet is an U-Net-like architecture. The encoder of SeResUNet contains a SeResNet to learns high-level semantic features and model the long-range dependencies among different channels of the learned feature maps. The decoder is the same as the vanilla U-Net. The encoder and decoder are connected by the skip connections for feature concatenation. We used the kidney and kidney tumor segmentation 2021 dataset to evaluate the proposed method. The dice, surface dice and tumor dice score of SeResUNet are 67.2%, 54.4%, 54.5%, respectively.

Keywords: Kidney tumor segmentation · Squeeze-and-excitation network · U-Net

1 Introduction

Kidney cancer is the malignant tumor with the highest mortality in the urinary system. Computed tomography (CT) imaging is the most common medical treatment for kidney cancer inspection and diagnosis. Segmenting kidneys and tumors from CT images is an important prerequisite for medical methods such as

© Springer Nature Switzerland AG 2022
N. Heller et al. (Eds.): KiTS 2021, LNCS 13168, pp. 71–79, 2022.
https://doi.org/10.1007/978-3-030-98385-7_10

pathological localization and radiotherapy planning. This is usually done manually by professional medical personnel or staff with relevant backgrounds. However, manual segmentation of kidney and kidney tumor from a large number of slices is time-consuming and suffers from human error. Automatic segmentation of the kidney and tumor can help doctors quickly locate the tumor and prepare for further surgical planning. The expansion of public databases has extremely promoted the segmentation of medical images. The kidney tumor segmentation challenge 2019 (KiTS19) [4] first released a public data set with kidney tumor annotations for the participants to develop automated segmentation approaches. KiTS19 provides 300 high-quality CT scan images of kidney cancer patients. Among them, 210 high-quality annotated CT scans are used for training, and 90 CT scans are used for algorithm testing. The KiTS19 Challenge has greatly promoted the segmentation of kidneys and kidney tumors.

Recently, deep learning-based methods have achieved impressive performance on medical image segmentation. Specifically, U-Net and its variants [1,8,11,14] are widely exploited for kidney and kidney tumor segmentation. For example, Yang et al. proposed a 3D full convolutional network combined with a pyramid pooling module (PPM) for kidney and kidney tumors segmentation, which can make full use of the 3D spatial contextual information to improve the segmentation of the kidney as well as the tumor lesion [12]. Abhinav Dhere et al. used the anatomical asymmetry of the kidney to define an effective kidney segmentation agent task through self-supervised learning [2]. Yu et al. proposed a framework named Crossbar-Net, which through vertical patches and horizontal patches to capture both the global and local appearance information of the kidney tumors, and cascade the horizontal sub-model with the vertical sub-model to segment the kidney and tumor [13]. Isensee et al. use the nnUnet for kidney and kidney tumor segmentation, which won the 1nd place in the kidney tumor segmentation challenge 2019 (KiTS2019) [7]. Hou et al. proposed a three-stage self-guided network to accurately segment kidney tumors. The first stage determines the rough position of the target, the second stage optimize, smooth kidney boundary and get the initial tumor segmentation result, the tumor refine net is proposed to optimize previous stage's tumor segmentation result in the third stage, which ranked the 2nd place in the KiTS19 [5]. Although these methods achieved favorable performance, the long-range dependencies of feature maps learned by convolutional neural network (CNN) are overlooked, which leaves room for further improvement.

Motivated by the squeeze-and-excitation network [6] to model long-range dependencies of the learned feature maps, in this paper, we propose a squeeze-and-excitation encoder-decoder network, named SeResUNet, for kidney and kidney tumor segmentation. Specifically, SeResUNet is an U-Net-like architecture including an encoder, a decoder, four skip connection paths. The encoder of SeResUNet contains a SeResNet to learns high-level semantic features and model the long-range dependencies among different channels of the learned feature maps. The decoder is the same as the vanilla U-Net. The encoder and decoder are connected by the skip connections for feature concatenation. We evaluated the proposed method on the 2021 kidney and kidney tumor segmentation challenge (KiTS21) dataset [4]. Experiment result shows that the dice, surface dice and tumor dice scores of SeResUNet are 67.2%, 54.4%, 54.5%, respectively.

2   Method

In this section, we detail our architecture for automated kidney and kidney tumor segmentation in CT images. First, We introduced the overall architecture in Sect. 2.1. In Sect. 2.2, the squeeze-and-excitation module are specified. Then, in Sect. 2.3, we present the deep supervision used in this work. Finally, the loss function is discussed in Sect. 2.4.

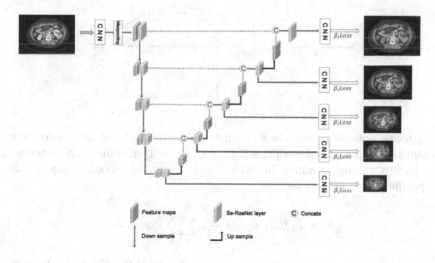

Fig. 1. Network architecture for segmentation

2.1   Architecture

The overview of our proposed framework is shown in Fig. 1. Our method is an encoder-decoder architecture. The encoder adopts ResNet50 [3] as backbone, including four residual blocks followed by maxpooling layers, to gradually aggregate high-level semantic information. In addition, we exploit the Squeeze-and-Excitation (SE) module in the encoder to model long-range dependencies of channel relation among the input feature maps. Specifically, SE module transform the feature maps into a channel descriptor, then recalibrate the input features themselves by channel-wise multiplication. The decoder up-samples the high-level feature map to obtain the segmentation map with size the same as the original image. Each convolution layer of decoder is of kernel size of 3×3, stride of 1, and padding of 1. In order to avoid the problem of vanishing gradient and to train the proposed network quickly, we introduce multi-level deep super-vision in the decoder, where deep supervision is performed on each layer of the decoder so that the shallow layer can be fully trained. After the four-layer up-sampling of the decoder, a segmentation map with the same size as the original image is obtained.

2.2 Squeeze-and-Excitation Module

We employ Squeeze-and-Excitation (SE) module [6] to capture channel-dependencies of the learned features. The structure of the SE module is depicted in Fig. 2.

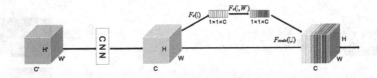

Fig. 2. Squeeze-and-Excitation module

Squeeze operation compresses each two-dimensional feature channel into a real number with a global receptive field, and the output dimension matches the input feature channel number. In short, it is to carry out global average pooling, the specific equation is as follows:

$$S_c = F_s(\omega_c) = \frac{1}{H \times W} \sum_{\alpha=1}^{H} \sum_{\beta=1}^{W} \omega(\alpha, \beta) \tag{1}$$

where ω_c represents the c_{th} feature map of size $H \times W$. After Eq. 1, the $H \times W \times C$ input is converted to $1 \times 1 \times C$, which represents the numerical distribution of the C feature maps in this layer, corresponding to the $F_s(\cdot)$ in Fig. 2.

Excitation is similar to the gate in the recurrent neural network, expresses the correlation between different feature channels by generating weights for each feature channel, the specific equation is as follows:

$$E_c = F_e(S, W) = \sigma(g(S, W)) = \sigma(W_2 \delta(W_1 S)) \tag{2}$$

where σ refers to the ReLU [9] function, W_1, W_2 is a fully connected layer with different parameters, used to fuse feature map information of different channels. The dimension of E obtained after Eq. 2 is $1 \times 1 \times C$, where C is the number of channels.

Recalibration operation multiply the excitation output E by the previous features, completing the recalibration of the original feature in the channel dimension. The specific equation is as follows:

$$F_{scale}(\omega_c, E_c) = \omega_c \cdot E_c \tag{3}$$

2.3 Deep Supervision

To avoid the vanishing gradient problem and quickly train the proposed network, we perform deep supervision in the decoder, as shown in the right of Fig. 1. Specifically, each layer of the decoder predicts a segmentation map for the calculation of the loss function. This is different from multi-task learning (MTL). MTL has different ground truths to calculate different losses, while there is only one ground truth for deep supervision. Different network layers calculate loss and sum them according to different coefficients. The weighted coefficients are set as 0.4, 0.3, 0.2, 0.05, 0.05, respectively. Since the sizes of feature map of each output layer are different, we down-sample the ground truth to the same size of the corresponding output segmentation map for loss calculation.

2.4 Loss Function

Loss function is used to estimate the degree of inconsistency between the predicted segmentation map $f(x)$ of the model and the ground truth Y. Considering the proportions of the volume of the kidney, tumor, cyst and background area are different, there is an imbalance in data distribution. Therefore, we use weighted cross-entropy (WCE) loss function to solve this problem. The specific definition is as follows:

$$\mathcal{L}_{wce}(\beta, P) = \sum_{i=0}^{S} \beta^i P_{GT}^i \log\left(p_{pred}^i\right) \qquad (4)$$

where S is the number of classes, specific for kidney, kidney tumor, kidney cyst and background. P_{GT}^i and p_{pred}^i are the probabilities of the i_{th} class of the ground truth and the prediction respectively, β^i is the weight of the i_{th} class. Here, the weights β^i are set to 1.0, 2.0, 4.0 and 4.0 for kidney, tumor, cyst and background respectively, in Eq. 4 according to the preliminary experiments.

3 Experiments

In this section, we illustrate the KiTS21 dataset on Sect. 3.1. Then, the evaluation metrics are presented on Sect. 3.2. Next, we describe the pre-process and post-process methods on Sect. 3.3. Finally, we specify the implementation details on Sect. 3.4.

3.1 Datasets

The CT scans used in this work come from The 2021 Kidney and Kidney Tumor Segmentation Challenge (KiTS21), which contains 300 complete data of kidneys, kidney tumors, kidney cysts and background labels. We randomly divided 60 data into one big category, divided into five in total, for five-fold cross-validation.

The data labels provided by the KiTS21 challenge are not exactly the same. Some data contains 4 complete labels, and some may only contain 3 of them. This also makes the training of our model difficult. Figure 3 shows some of them.

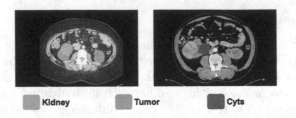

Fig. 3. Illustration of the annotations in the KiTS21 dataset.

3.2 Metrics

This article employs Dice and surface Dice similarity coefficient (surface DSC)[10] to evaluate our segmentation results. Dice is one of the most commonly used evaluation indicators. Specifically, we use kidneys, kidney tumors, and cysts as the foreground, everything else as background to calculate Dice scores. In medical images, the voxel spacing is usually unequal, and the calculation of surface voxels usually leads to larger errors. The surface Dice used in this article is given an allowable error distance, and the surface within this error range is regarded as the overlapping part, the surface overlap dice value of the ground truth mask and the predict mask is calculated.

3.3 Pre- and Post-processing

We consider using 2D U-Net to complete our experiments. KiTS21 challenge provides 3D dataset, we first convert voxels into slice data, highlight organ and tumor features through threshold processing and threshold normalization, enhance the data by flipping, random cropping, and random translation.

In the post-processing part, considering the influence of noise, we perform the operation of the largest connected domain on the data to eliminate the noise.

3.4 Implementation Details

First, we initialize the model parameters, set the training epoch to 20, the initial learning rate to 10^{-5}, batchsize is set to 8, and use the Adam optimizer. The proposed network was implemented in python using Pytorch (v1.5.1) framework in the backend. All training and testing experiments are run on a workstation with an NVIDIA GeForce GTX 2080Ti with 11G GPU memory.

4   Result

We evaluated our model on KiTS21 dataset through five-fold cross-validation. The results were obtained by averaging the best performance of each fold. Table 1 reports the quantitative results of the proposed SeResUNet and the most commonly used segmentation methods. The dice of kidney, masses and tumor we test are 91.60%, 58.80%, 54.16% respectively, and the surface dice are 84.62%, 37.91%, 37.59% respectively, and the result of dice, surface dice and tumor dice in KiTS21 are 67.2%, 54.4%, 54.5% respectively. In addition, the segmentation results we test are shown in Fig. 4.

Table 1. Dice score (mean) and surface dice of the proposed method on 5-fold cross validation.

Method	Kidney (Dice)	Masses (Dice)	Tumor (Dice)	Kidney (SD)	Masses (SD)	Tumor (SD)
2D U-Net	0.9132	0.3769	0.3712	0.8425	0.2618	0.2573
SeResUNet18	0.8801	0.3923	0.3556	0.7952	0.2677	0.2370
SeResUNet50+D	0.9144	0.5797	0.5274	0.8396	0.4187	0.3741
SeResUNet18+D	0.9160	0.5880	0.5416	0.8462	0.3791	0.3759

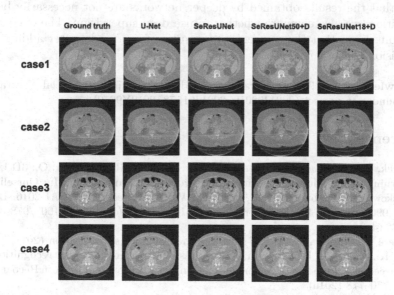

Fig. 4. Qualitative comparison of segmentation results for KiTS21 dataset test:Ground truth, U-Net, SeResUNet, SeResUNet50 with deep supervision (SeResUNet50+D), and proposed SeResUNet18 with deep supervision (SeResUNet18+D). Color coding: red,Kidney; green,Tumor; blue,Cyts. (Color figure online)

5 Discussion and Conclusion

In this work, we proposed a novel segmentation network called SeResUNet to deal with the kidney and tumor segmentation task. First, we adopt the encoder-decoder architecture like U-Net, and use ResNet to deepen the network in the encoder. At the same time, in order to avoid deep network degradation problems and speed up the convergence, we add multi-level deep supervision to the decoder. In addition, we noticed the importance of different channels and introduced Squeeze-and-Excitation module, which automatically obtains the weight of each feature channel through learning, and then highlights useful features based on this weight and suppresses features that are not useful for the current task. Finally, the weight cross-entropy loss function is used to solve the problem of data imbalance. Through the evaluation on the KiTS21 dataset, it can be seen that the model we proposed has a stronger ability in the kidney and its tumor segmentation.

It can be seen from Table 1 that our network is improved by 0.28%, 21.1% and 16.88% respectively compared with the classic network 2D U-Net. The segmentation results of the kidney have not improved much, but the segmentation results of tumors and cysts have improved greatly, indicating that our model performs better in the subtle parts. Compared with Se-ResUNet18 and Se-ResUNet50, kidney, masses and tumor are increased by 0.16%, 0.83%, 1.42% respectively, it shows that the results obtained by deeper networks are not necessarily better. In addition, we also compare whether to use deep supervision. The experimental result shows that the segmentation results of our model after adding deep supervision will be much better.

Acknowledgment. This work was supported by the Fujian Provincial Natural Science Foundation project (2021J02019, 2021J01578, 2019Y9070).

References

1. Çiçek, Ö., Abdulkadir, A., Lienkamp, S.S., Brox, T., Ronneberger, O.: 3D U-Net: learning dense volumetric segmentation from sparse annotation. In: Ourselin, S., Joskowicz, L., Sabuncu, M.R., Unal, G., Wells, W. (eds.) MICCAI 2016. LNCS, vol. 9901, pp. 424–432. Springer, Cham (2016). https://doi.org/10.1007/978-3-319-46723-8_49
2. Dhere, A., Sivaswamy, J.: Self-supervised learning for segmentation (2021)
3. He, K., Zhang, X., Ren, S., Sun, J.: Deep residual learning for image recognition. In: Proceedings of the IEEE Conference on Computer Vision And Pattern Recognition, pp. 770–778 (2016)
4. Heller, N., et al.: The kits19 challenge data: 300 kidney tumor cases with clinical context, CT semantic segmentations, and surgical outcomes. arXiv preprint: arXiv:1904.00445 (2019)
5. Hou, X., et al.: A triple-stage self-guided network for kidney tumor segmentation. In: 2020 IEEE 17th International Symposium on Biomedical Imaging (ISBI), pp. 341–344. IEEE (2020)
6. Hu, J., Shen, L., Sun, G.: Squeeze-and-excitation networks. In: Proceedings of the IEEE Conference on Computer Vision and Pattern Recognition, pp. 7132–7141 (2018)
7. Isensee, F., Maier-Hein, K.H.: An attempt at beating the 3D U-Net (2019)
8. Milletari, F., Navab, N., Ahmadi, S.A.: V-net: fully convolutional neural networks for volumetric medical image segmentation. In: 2016 Fourth International Conference on 3D Vision (3DV), pp. 565–571. IEEE (2016)
9. Nair, V., Hinton, G.E.: Rectified linear units improve restricted Boltzmann machines. In: ICML (2010)

10. Nikolov, S., et al.: Deep learning to achieve clinically applicable segmentation of head and neck anatomy for radiotherapy. arXiv preprint: arXiv:1809.04430 (2018)
11. Ronneberger, O., Fischer, P., Brox, T.: U-Net: convolutional networks for biomedical image segmentation. In: Navab, N., Hornegger, J., Wells, W.M., Frangi, A.F. (eds.) MICCAI 2015. LNCS, vol. 9351, pp. 234–241. Springer, Cham (2015). https://doi.org/10.1007/978-3-319-24574-4_28
12. Yang, G., et al.: Automatic kidney segmentation in CT images based on multi-atlas image registration. In: 2014 36th Annual International Conference of the IEEE Engineering in Medicine and Biology Society, pp. 5538–5541. IEEE (2014)
13. Yu, Q., Shi, Y., Sun, J., Gao, Y., Zhu, J., Dai, Y.: Crossbar-net: a novel convolutional neural network for kidney tumor segmentation in CT images. IEEE Trans. Image Process. 28(8), 4060–4074 (2019)
14. Zhou, Z., Rahman Siddiquee, M.M., Tajbakhsh, N., Liang, J.: U Net++: a nested U-Net architecture for medical image segmentation. In: Stoyanov, D., et al. (eds.) DLMIA/ML-CDS -2018. LNCS, vol. 11045, pp. 3–11. Springer, Cham (2018). https://doi.org/10.1007/978-3-030-00889-5_1

A Two-Stage Cascaded Deep Neural Network with Multi-decoding Paths for Kidney Tumor Segmentation

Tian He, Zhen Zhang, Chenhao Pei, and Liqin Huang$^{(\boxtimes)}$

College of Physics and Information Engineering, Fuzhou University, Fuzhou, China
hlq@fzu.edu.cn

Abstract. Kidney cancer is aggressive cancer that accounts for a large proportion of adult malignancies. Computed tomography (CT) imaging is an effective tool for kidney cancer diagnosis. Automatic and accurate kidney and kidney tumor segmentation in CT scans is crucial for treatment and surgery planning. However, kidney tumors and cysts have various morphologies, with blurred edges and unpredictable positions. Therefore, precise segmentation of tumors and cysts faces a huge challenge. Consider these difficulties, we propose a cascaded deep neural network, which first accurately locate the kidney area through 2D U-Net, and then segment kidneys, kidney tumors, renal cysts through Multi-decoding Segmentation Network (MDS-Net) from the ROI of the kidney. We evaluated our method on the 2021 Kidney and Kidney Tumor Segmentation Challenge (KiTS21) dataset. The method achieved Dice score, Surface Dice and Tumor Dice of 69.4%, 56.9% and 51.9% respectively, in the test cases. The model of cascade network proposed in this paper has a promising application prospect in kidney cancer diagnosis.

Keywords: Kidney · Kidney tumor segmentation · Cascaded deep neural network · Multi-decoding

1 Introduction

Renal cell carcinoma (RCC) is a malignant tumor formed by the malignant transformation of epithelial cells in different parts of the renal tubule [6]. Its incidence accounts for 80% to 90% of adult renal malignant tumors, and the prevalence of men is higher than that of women [1,4,5]. The incidence of kidney cancer is closely related to genetics, smoking, obesity, hypertension, and antihypertensive therapy, which is second only to prostate cancer and bladder cancers among tumors of the urinary system. Accurate segmentation of tumors from 3D CT remains a challenging task due to the unpredictable shape and location of tumors in the patient, as well as the confusion of textures and boundaries [9,19].

The traditional method of manually segmenting tumors is not only time-consuming and laborious, but also has the problem of inconsistent results during

© Springer Nature Switzerland AG 2022
N. Heller et al. (Eds.): KiTS 2021, LNCS 13168, pp. 80–89, 2022.
https://doi.org/10.1007/978-3-030-98385-7_11

segmentation by senior doctors, which leads to unsatisfactory results in clinical applications [6,13,17]. Therefore, computer-assisted kidney tumor segmentation methods have attracted much attention. In recent years, deep learning has penetrated into various application fields, and its performance in many fields such as image detection, classification, and segmentation has exceeded the most advanced level [7]. Among current CNN-based methods, the popular U-Net [16] and 3D U-Net [3] architecture have exhibited promising results in medical image segmentation tasks [10], such as pancreas segmentation [14], prostate segmentation [18] and brain segmentation [15]. Although 3D Fully Convolutional Network (FCN) [12] segmentation performance is higher than 2D FCN, it requires greater memory consumption. Zhang et al. [20] proposed a cascaded framework network for automatic segmentation of kidneys and tumors, which alleviates the problem of inaccurate segmentation caused by insufficient network depth due to excessive memory consumption. With extremely limited data, a cascaded 3D U-Net with a active learning function can improve training efficiency and reduce labeling work [8].

Recently, Li et al. [10] proposed a 3D U-Net based on memory efficiency and non-local context guidance, which captures the global context through a non-local context guidance mechanism and fully utilize long-distance dependence in the feature selection process. In the 3D U-Net, this method complements high-level semantic information with spatial information through a layer skip connection between the encoder and the decoder, and finally realizes the precise segmentation of the kidney and the tumor.

In this work, we develop a fully automatic cascaded segmentation network with multi-decoding paths. The kidney area is first located through 2D U-Net. The area is cropped according to the region of interest located in the first stage and input it into MDS-Net to accurately segment the kidney, renal tumor and renal cyst. General, the contributions of our work can be summarized in the following three aspects:

1. We develop a two-stage cascaded segmentation network with multi-decoding paths and evaluate it on the 2021 Kidney and Kidney Tumor Segmentation Challenge (KiTS21) dataset.
2. We propose a fusion module based on global context (GC) [2], which can realize the attention to channel and spatial context to achieve noise suppression and enhancement of useful information.
3. We present a regional constraint loss function, which is used to measure the constraint relationship of impassable regions.

2 Methods

Figure 1 shows the two-stage cascaded deep neural network for kidney tumor segmentation. First input the pre-processed image to locate the kidney through 2D U-Net to obtain an accurate kidney area, then use the kidney area as the bounding box of the original CT, cropping to get the input image, and train the MDS-Net to segment kidney, tumors and cysts.

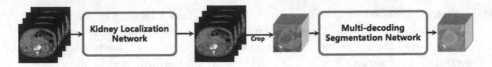

Fig. 1. The two-stage cascaded segmentation framework

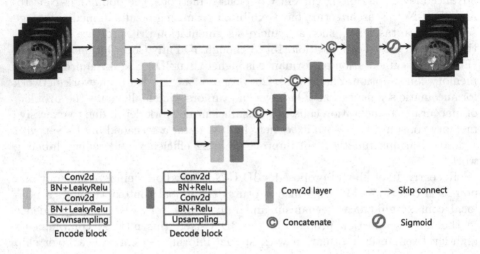

Fig. 2. The kidney localization network (U-Net)

2.1 Kidney Localization Network

For the localization of the kidney, we trained a 2D U-Net for kidney segmentation and localization. As shown in Fig. 2, the encoding path composed of four encoder blocks, and each block is composed of 2D convolution, Batchnorm, LeakyReLU and downsampling. On the decoding path, each decoding block is composed of 2D convolution, Batchnorm, ReLU, and upsampling. After upsampling on the last layer, the image undergoes a 3×3 convolution. The input of the network is a 256×256 image, and the output is divided into the background and the kidney through a Sigmoid function. The loss function used is Dice loss

$$\mathcal{L}oss_{KI} = 1 - DSC(L_{KI}, \hat{L}_{KI}), \tag{1}$$

where $DSC(A, B)$ calculates the Dice similarity coefficient of A and B, L_{KI} and $\hat{L}_{KI}$ are the corresponding gold standard and predicted label of whole regions including the kidneys, tumors, and cysts. For the segmentation results, we performed connected components analysis and selected the largest two connected component to locate the kidney.

2.2 Multi-decoding Segmentation Network

Multi-decoding segmentation network (MDS-Net) is designed to segment normal kidneys, kidney tumors, and kidney cysts. The Fig. 3 shows the design of

MDS-Net, which consists of an encoding path, three decoding paths, and a fusion prediction branch.

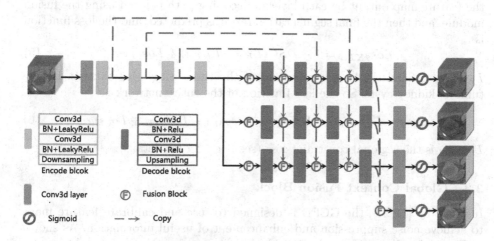

Fig. 3. Multi-decoding segmentation network

The image patch obtained by the first stage cropping is input to the encoding path for feature extraction, and three segmentation results ($\hat{L}_{KI}$, $\hat{L}_{MA}$, $\hat{L}_{TU}$) are obtained by the three decoding paths. Fusion of the features obtained by the three decoding paths to obtain the final segmentation result $\hat{L}_{KTC}$.

Due to the imbalance of the segmentation classes, e.g. the cyst does not exist in any cases or only occupies a small area, which makes the network difficult to train. Therefore, we set the three regions of the target segmentation as KI is the entire kidney region, including normal kidney, tumor and cyst, MA is kidney masses that include tumors and cysts region, and TU is the region of tumors only. By decomposing the original multi-label segmentation task into these three single-label segmentation tasks, the impact of category imbalance is reduced.

In the encoding path, feature extraction is performed by a convolutional layer and four encode blocks, and each encode block is composed of a 3D convolutional layer, Batchnorm, LeakyRelu, and downsampling. The future map obtained after downsampling is used as the input of the next module, and is also input into the decoding path through a skip connection. Each decoding branch is composed of a feature global context fusion block (see Sect. 2.3 for details), decoder, and a convolutional layer. The decode block is similar to the encode block, while the last layer is upsampling. The fusion block fuses and corrects the output feature maps from the previous layer and skip connection. The features obtained by the three decoding paths are output through the Sigmoid layer to obtain ($\hat{L}_{KI}$, $\hat{L}_{MA}$, $\hat{L}_{TU}$) separately, the corresponding loss functions are

$$\mathcal{L}oss_{KI} = 1 - DSC(L_{KI}, \hat{L}_{KI}), \tag{2}$$

$$\mathcal{L}oss_{MA} = 1 - DSC(L_{MA}, \hat{L}_{MA}), \tag{3}$$

$$Loss_{TU} = 1 - DSC(L_{TU}, \hat{L}_{TU}), \tag{4}$$

where (L_{KI}, L_{MA}, L_{TU}) is the ground truth of region (KI, MA, TU). Finally, the feature map output by each layer of decoding path is fused using the fusion module, and then the final segmentation result is predicted, and the loss function is

$$Loss_{KTC} = 1 - DSC(L_{KTC}, \hat{L}_{KTC}) + Loss_{RC}, \tag{5}$$

L_{KTC} is defined as the ground stand of the three categories kidneys, kidney tumors, kidney cysts. So the loss function of the entire network is

$$Loss = Loss_{KTC} + Loss_{KI} + Loss_{MA} + Loss_{TU} + Loss_{RC}, \tag{6}$$

$Loss_{RC}$ is the regional constraint loss (see Sect. 2.4 for details).

2.3　Global Context Fusion Block

Inspired by [11, 21], the GCFB is designed to fuse and calibrate feature maps to achieve noise suppression and enhancement of useful information. As shown in Fig. 4, the Global Context (GC) block that combines Non-local and Squeeze-and-Excitation (SE) block is used to calibrate the features, which can realize the attention to the channel and spatial context, and obtain the fused feature map through the convolutional layer and ReLU.

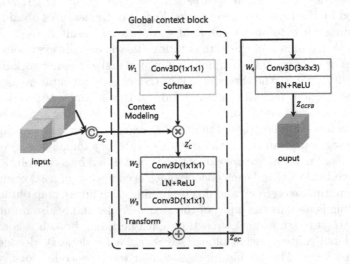

Fig. 4. Global context fusion block

In GCFB, first concat the input on the channel to get Z_C. The GC block can be defined as:

$$Z'_C = Z_C \otimes Softmax(W_1 Z_C), \tag{7}$$

$$Z_{GC} = Z_C + W_3 ReLU(LN(W_2 Z'_C)), \tag{8}$$

W_* is the parameters of the convolution layers. Therefore, the final output of the GCFB is

$$Z_{GCFB} = ReLU(BN(W_4 Z_{GC})). \qquad (9)$$

2.4 Regional Constraint Loss Function

As shown in Fig. 5, there are regional relationships in different regions of the kidney, the region of KI contains MA, TU is in MA. In order to achieve the constraint of this relationship, the overlap degree of different regions [11] is calculated to measure whether the constraint relationship between the regions is satisfied. The regional constraint loss function is

$$\mathcal{L}oss_{RC} = 1 - \frac{1}{2} \left(\frac{\sum\limits_{x \in \Omega} \hat{L}_{KI}(x) \cdot \hat{L}_{MA}(x)}{\sum\limits_{x \in \Omega} \hat{L}_{MA}(x)} + \frac{\sum\limits_{x \in \Omega} \hat{L}_{MA}(x) \cdot \hat{L}_{TU}(x)}{\sum\limits_{x \in \Omega} \hat{L}_{TU}(x)} \right), \qquad (10)$$

where $(\hat{L}_{KI}, \hat{L}_{MA}, \hat{L}_{TU})$ is predicted result from three decoding path from MDS-Net (see Fig. 3), the Ω is the common spatial space.

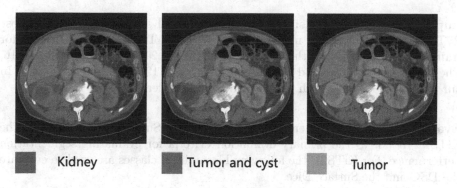

Fig. 5. Regional constraint of kidney

3   Experimental Results

3.1   Dataset

The KiTS21 Challenge provides contrast-enhanced CT scans and annotation data from 300 patients who underwent partial or radical nephrectomy for suspected renal malignant tumors at M Health Fairview or Cleveland Clinic Medical Center between 2010 and 2020, which provides us with three annotation data, and we finally chose the voxel-wise majority voting aggregations for training and validation. The size, shape and density of the tumor are various in different CT scans. Moreover, only a few images of existing cysts. The annotation work is completed by experienced experts, trainees and laid-off workers together, and the annotated data is used as the ground truth of the training. Since only the training set data was provided, we randomly divided the data of the 300 cases into 5 pieces, each with 60 cases. One of the 5 pieces was selected in turn as the verification set and the other 4 pieces as the training set.

3.2   Implementation Details

Data Processing: Before training our cascade model, we first performed a crop slice operation to make all volume slices the same thickness to reduce GPU memory consumption and training time. For the first network, we input 2D axial slices, which are obtained by extracting slices from the original 3D CT along the z-axis and adjusting the size from 512×512 to 256×256. For the second network MDS-Net, according to the maximum rectangular frame range of the region of each kidney, the size of the block of the region of interest extracted is $128 \times 128 \times 128$. Then, we truncated the image intensity values of all images to the range of $[-100, 500]$ HU to remove the fat area around the kidney and remove irrelevant details.

Implementation Details: As an experimental environment, we choose PyTorch to implement our model and use NVIDIA Tesla P100 16GB GPU for training. The input size of the first network is 256×256 with a batch size of 16, The input size of the second network is $128 \times 128 \times 128$ with batch size of 4. In our model, we set the epoch to 200, and the initial learning rate is 1×10^{-2}.

Evaluation Metrics: We employ the DSC and the Surface Dice provided in the KiTS21 toolkit as the primary evaluation criteria for evaluating segmentation performance. For KiTS21, the following hierarchical classes are used to evaluate the DSC and the Surface Dice.

- Kidney and Masses (KI): Kidney + Tumor + Cyst,
- Kidney Mass (MA): Tumor + Cyst,
- Tumor (TU): Tumor only.

3.3   Results

To evaluate the effectiveness of our method, we compared our network with other state-of-the-art methods, including 2D U-Net and 3D U-Net. Furthermore, to explore the advantage of GCFB and $Loss_{RC}$, we also compared our method with the method without GCFB and $Loss_{RC}$. We perform visual and statistical comparisons under the same data set and data parameters. In addition, in order to explore the advantages of our method, we use the evaluation index Dice coefficient and the Surface Dice to verify our method on the KiTS 21 data set. We compared the results of our method with four different methods:

Table 1. Dice score and Surface Dice of the proposed method and other baseline methods on the validation set.

Method	Dice (%)			Surface dice		
	KI	MA	TU	KI	MA	TU
2D U-Net	90.78	43.61	44.86	83.30	28.21	29.51
3D U-Net	91.59	59.29	57.71	83.76	40.55	39.95
Ours (wo GCFB)	92.55	63.23	58.63	83.84	42.11	39.11
Ours (wo $Loss_{RC}$)	93.12	67.91	62.78	85.75	**46.37**	42.25
Ours	**93.48**	**68.32**	**64.26**	**86.29**	45.90	**43.34**

From Table 1, compared to 2D U-Net and 3D U-Net, our proposed methods performed better both in Dice and Surface Dice. It also shows that our proposed methods with GCFB and $Loss_{RC}$ can achieve better results than our methods without GCFB or $Loss_{RC}$. In addition, Fig. 6 shows the visualization results of different methods on the validation set. In Fig. 6, our method is more effective than other methods in easier cases and challenging cases. We finally used the trained cascaded model to perform the prediction on the test cases. We obtained Dice score, Surface Dice and Tumor Dice of 69.4%, 56.9% and 51.9% respectively.

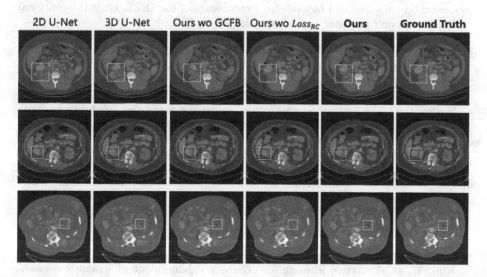

| 2D U-Net | 3D U-Net | Ours wo GCFB | Ours wo $Loss_{RC}$ | **Ours** | **Ground Truth** |

Fig. 6. Visualization results of different methods on the validation set. The segmentation contained in the yellow dashed box is our concern. (Color figure online)

4   Conclusion

In this work, we proposed a novel two-stage cascade and multi-decoding method for kidney segmentation. We utilized U-Net to achieve the location and extraction of the region of kidney, and then designed MDS-Net for the final segmentation. For MDS-Net, we developed a segmentation network with multiple decoders, combined the features of the three decoding paths with GCFB, and obtained the final segmentation result. Besides, we presented a regional constraint loss function to predict the segmentation result with more reality. It has been evaluated on the dataset from KITS 2021. Experimental results show that this method obtains good segmentation results on kidney tumors.

Acknowledgment. This work was financed by Fujian Provincial Natural Science Foundation project (Grant No. 2021J02019, 2021J01578, 2019Y9070), Fuzhou Science and Technology Project (2020-GX-17).

References

1. Bray, F., Ferlay, J., Soerjomataram, I., Siegel, R.L., Torre, L.A., Jemal, A.: Global cancer statistics 2018: GLOBOCAN estimates of incidence and mortality worldwide for 36 cancers in 185 countries. CA: Cancer J. Clin. **68**(6), 394–424 (2018)
2. Cao, Y., Xu, J., Lin, S., Wei, F., Hu, H.: GCNet: non-local networks meet squeeze-excitation networks and beyond. In: Proceedings of the IEEE/CVF International Conference on Computer Vision Workshops (2019)
3. Çiçek, Ö., Abdulkadir, A., Lienkamp, S.S., Brox, T., Ronneberger, O.: 3D U-Net: learning dense volumetric segmentation from sparse annotation. In: Ourselin, S., Joskowicz, L., Sabuncu, M.R., Unal, G., Wells, W. (eds.) MICCAI 2016. LNCS, vol. 9901, pp. 424–432. Springer, Cham (2016). https://doi.org/10.1007/978-3-319-46723-8_49
4. Ferlay, J., et al.: Estimating the global cancer incidence and mortality in 2018: Globocan sources and methods. Int. J. Cancer **144**(8), 1941–1953 (2019)
5. Ferlay, J., Shin, H.R., Bray, F., Forman, D., Mathers, C., Parkin, D.M.: Estimates of worldwide burden of cancer in 2008: Globocan 2008. Int. J. Cancer **127**(12), 2893–2917 (2010)
6. Guo, J., Zeng, W., Yu, S., Xiao, J.: Rau-net: U-net model based on residual and attention for kidney and kidney tumor segmentation. In: 2021 IEEE International Conference on Consumer Electronics and Computer Engineering (ICCECE), pp. 353–356. IEEE (2021)
7. Havaei, M., et al.: Brain tumor segmentation with deep neural networks. Med. Image Anal. **35**, 18–31 (2017)
8. Kim, T., et al.: Active learning for accuracy enhancement of semantic segmentation with CNN-corrected label curations: evaluation on kidney segmentation in abdominal CT. Sci. Rep. **10**(1), 1–7 (2020)
9. Li, X., Liu, L., Heng, P.A.: H-DenseUNet for kidney and tumor segmentation from CT scans (2019)
10. Li, Z., Pan, J., Wu, H., Wen, Z., Qin, J.: Memory-efficient automatic kidney and tumor segmentation based on non-local context guided 3D U-Net. In: Martel, A.L., et al. (eds.) MICCAI 2020. LNCS, vol. 12264, pp. 197–206. Springer, Cham (2020). https://doi.org/10.1007/978-3-030-59719-1_20

11. Liu, C., et al.: Brain tumor segmentation network using attention-based fusion and spatial relationship constraint. arXiv preprint arXiv:2010.15647 (2020)
12. Long, J., Shelhamer, E., Darrell, T.: Fully convolutional networks for semantic segmentation. In: Proceedings of the IEEE Conference on Computer Vision and Pattern Recognition, pp. 3431–3440 (2015)
13. Mu, G., Lin, Z., Han, M., Yao, G., Gao, Y.: Segmentation of kidney tumor by multi-resolution VB-nets (2019)
14. Oktay, O., et al.: Attention U-Net: learning where to look for the pancreas. arXiv preprint arXiv:1804.03999 (2018)
15. Rickmann, A.-M., Roy, A.G., Sarasua, I., Navab, N., Wachinger, C.: 'Project & excite' modules for segmentation of volumetric medical scans. In: Shen, D., et al. (eds.) MICCAI 2019. LNCS, vol. 11765, pp. 39–47. Springer, Cham (2019). https://doi.org/10.1007/978-3-030-32245-8_5
16. Ronneberger, O., Fischer, P., Brox, T.: U-Net: convolutional networks for biomedical image segmentation. In: Navab, N., Hornegger, J., Wells, W.M., Frangi, A.F. (eds.) MICCAI 2015. LNCS, vol. 9351, pp. 234–241. Springer, Cham (2015). https://doi.org/10.1007/978-3-319-24574-4_28
17. Taha, A., Lo, P., Li, J., Zhao, T.: Kid-Net: convolution networks for kidney vessels segmentation from CT-volumes. In: Frangi, A.F., Schnabel, J.A., Davatzikos, C., Alberola-López, C., Fichtinger, G. (eds.) MICCAI 2018. LNCS, vol. 11073, pp. 463–471. Springer, Cham (2018). https://doi.org/10.1007/978-3-030-00937-3_53
18. Yu, L., Yang, X., Chen, H., Qin, J., Heng, P.A.: Volumetric convnets with mixed residual connections for automated prostate segmentation from 3D MR images. In: Thirty-First AAAI Conference on Artificial Intelligence (2017)
19. Yu, Q., Shi, Y., Sun, J., Gao, Y., Zhu, J., Dai, Y.: Crossbar-net: a novel convolutional neural network for kidney tumor segmentation in CT images. IEEE Trans. Image Process. **28**(8), 4060–4074 (2019)
20. Zhang, Y., et al.: Cascaded volumetric convolutional network for kidney tumor segmentation from CT volumes. arXiv preprint arXiv:1910.02235 (2019)
21. Zhang, Z., et al.: Multi-modality pathology segmentation framework: application to cardiac magnetic resonance images. In: Zhuang, X., Li, L. (eds.) MyoPS 2020. LNCS, vol. 12554, pp. 37–48. Springer, Cham (2020). https://doi.org/10.1007/978-3-030-65651-5_4

Mixup Augmentation for Kidney and Kidney Tumor Segmentation

Matej Gazda[1]([✉]), Peter Bugata[1], Jakub Gazda[2], David Hubacek[3], David Jozef Hresko[1], and Peter Drotar[1]

[1] IISlab, Technical University of Kosice, Kosice, Slovakia
matej.gazda@tuke.sk
[2] 2nd Department of Internal Medicine, Pavol Jozef Safarik University and Louis Pasteur University Hospital, Kosice, Slovakia
[3] Department of Radiodiagnostics and Medical Imaging, Louis Pasteur University Hospital, Kosice, Slovakia

Abstract. Abdominal computed tomography is frequently used to non-invasively map local conditions and to detect any benign or malign masses. However, ill-defined borders of malign objects, fuzzy texture, and time pressure in fact, make accurate segmentation in clinical settings a challenging task. In this paper, we propose a two-stage deep learning architecture for kidney and kidney masses segmentation, denoted as convolutional computer tomography network (CCTNet). The first stage locates volume bounding box containing both kidneys. The second stage performs the segmentation of kidney, kidney tumors and cysts. In the first stage, we use a pre-trained 3D low resolution nnU-Net. In the second stage, we employ a mixup augmentation to improve segmentation performance of the second 3D full resolution nnU-Net. The obtained results indicate that CCTNet can provide improved segmentation of kidney, kidney tumor and cyst.

Keywords: nnU-Net · Mixup · Kidney · CT · Segmentation

1 Introduction

The kidney cancer is one of the ten most common cancers, and it is the third most common genitourinary malignancy [3] with high mortality. Kidney cysts, although being benign, are also a cause for concern due to their potential for malign transformation. Thus, it is essential to maximize the rate of their diagnosis, especially in early stages, when curative treatment is still possible. Medical imaging, such as MRI, CT and US, plays crucial role in detecting both kidney cyst and kidney cancer. Nowadays, kidney cancer cases are frequently found only incidentally during the regular medical image evaluation [4].

The incidence of renal cell carcinoma (RCC; one of the most common kidney cancers) is continuously growing and, since the 1990-ies, the incidence has doubled in developed world [4]. This may be attributed to advances in medical imaging technology, which provide more precise and detailed images than ever

© Springer Nature Switzerland AG 2022
N. Heller et al. (Eds.): KiTS 2021, LNCS 13168, pp. 90–97, 2022.
https://doi.org/10.1007/978-3-030-98385-7_12

before, and to a more frequent employment of imaging techniques in everyday clinical practice. RCC currently accounts for more than 400,000 new cases world wide every year [4]. It affects mainly the population of those older than 60 years and thus, with the aging world population, the number of patients is expected to increase even further.

Early diagnosis and subsequent treatment of kidney cancer is critical because it improves both health-related quality of life and overall survival. Development of computer-assisted diagnosis is important since it can help to automate MRI/CT/US evaluation, release some workload burden from radiologists, increase the number of evaluated images and in turn increase the proportion of RCC patients diagnosed in the very early stages, when curative treatment may be possible.

Currently, state-of-the-art models for semantic segmentation are based on U-net architecture, V-net architecture or their 3D derivatives. These architectures were also backbone for the recent Kidney and Kidney Tumor Segmentation Challenge 2019 (KITS 2019) [1]. The best performing models in final results were based on U-net and V-net architectures. The ultimate winner of KITS 2019 challenge nnU-net [2] proved itself also on several other segmentation challenges and currently represents gold standard in medical image semantic segmentation.

In this paper, we build upon the nnU-net and propose to use mixup [6] augmentation that was shown to improve the generalization of the state-of-the-art neural network architectures. The results obtained on published KITS2019 dataset indicate that mixup can be with advantage used also in segmentation of 3D CT images. We achieved improved performance on all evaluated classes.

In the next section, we provide the description of the proposed architecture together with the details of network training and validation. Then, the results of the proposed network and the baseline are given, followed by discussion.

2   Methods

Previous results indicate [2] that most of the performance gains are not network architecture dependent, but rather can be obtained by careful tuning of the network parameters. Another part of the machine learning pipeline that yields performance improvements is the data pre-processing and post-processing. As a such we did not attempt to find the architecture modifications to boost the performance but rather try to provide better condition for network training through mixup augmentation [6].

2.1   Training and Validation Data

Our submission made use of the official KITS21 training set alone and we used majority voting aggregation for multiple segmentation annotations. The public training dataset includes 3D CT scans of 300 of patients who underwent partial or radical nephrectomy for renal malignancy between years 2010 and 2020. The validation dataset used to evaluate performance consisted of scans of 100 patients. A review of these cases was conducted to identify all patients who had undergone a contrast-enhanced preoperative CT scan that includes the entirety of all kidneys. Gathered samples were annotated though extensive process, where annotation team was placed into three categories based on the level of knowledge. Their primary goal was to identify three semantic classes: kidney, tumor and cyst.

2.2   Preprocessing

The data transformation, re-sampling, and normalisation were handled by the nnU-Net configuration. As described in [2], this means for anisotropic data that image resampling strategy in place was third order spline interpolation for in-plane and nearest neighbours for out-of-plane. The normalisation was global dataset percentile clipping and z-score with global foreground mean and s.d. The clipping of HU values was handled by nnU-net default setting (0.5 and 99.5 percentile).

Since wrong labels can limit the quality of predictions learned by deep neural networks [5], we analysed true labels provided by the challenge organizers. Three cases were modified after the consultation with radiologist. In the case of 084, the mass structure attached to the left kidney is labeled as kidney. We cleared the kidney label, however we did not add any label for the mass, since it was difficult to decide whether this is a cyst or a tumor. The another modified case, 277, was more tricky. The labels provided by the organisers denote the cyst and tumor in the left kidney. However, closer inspection shows that this is in fact a single structure. In this case, we replaced the cyst label by the tumor label. Finally, in the third modified case - 299, we corrected the label for the left kidney. Axial view showed some nodule growing out of the left kidney, that is labeled as kidney, but has different density and does not fit to the kidney shape. We removed the kidney label for this nodule.

2.3   Proposed Method

The proposed classifier network consists of two cascade connected nnU-Net networks as depicted in Fig. 1. In the first stage we employ the nnU-Net to locate kidney in 3D CT image and crop volume containing kidneys. This step has two goals. To reduce computational cost in the second stage, and to eliminate possible erroneous predictions (e.g. cyst lying in some distant areas of CT image). Since the segmentation achieved by nnU-Net in KITS 2019 challenge was sufficiently high, we take advantage of the nnU-Net pre-trained on data from KITS 2019 challenge. To avoid some errors on the borders, we add 60 voxels in every direction to the cropped volume containing the kidneys.

To train nnU-Net in the second stage we use mixup [6] augmentation. Mixup is a data-agnostic augmentation routine that constructs virtual training examples by convex combinations of pairs of training examples and their labels according to the rule [6]

$$x_{mixup} = \lambda x_i + (1 - \lambda)x_j, \tag{1}$$

$$y_{mixup} = \lambda y_i + (1 - \lambda)y_j, \tag{2}$$

where (x_i, y_i) and (x_j, y_j) are two random training data samples, and $\lambda \in [0, 1]$. The λ is distributed according Beta distribution $\lambda \sim \beta(\alpha, \alpha)$, where $\alpha \in (0, \infty)$.

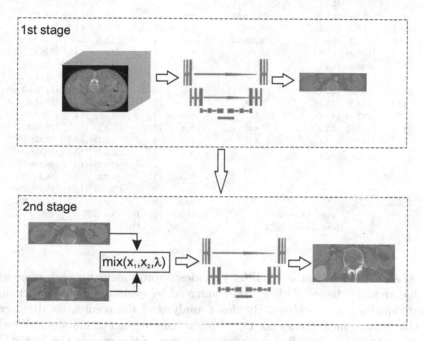

Fig. 1. Proposed two stage approach (CCTNet).

The networks parameters are automatically configured by nnU-Net, meaning that as a loss function a combination of Dice and Cross-entropy loss is used, and SGD with Nesterov momentum is used for network training optimization. To validate our approach, we used 5-fold cross-validation.

3   Results

In order to objectively asses the performance of the proposed approaches, three Hierarchical Evaluation Classes (HECs) were introduced : kidney and masses, kidney mass, and tumor. Kidney and masses HEC includes semantic classes kidney, tumor, and cyst. Kidney mass HEC covers only tumor and cyst. Finally, tumor HEC is the same as the semantic class for tumor.

To train the final model, we used batch size equal to four and the mixup augmentation with $\lambda \sim \beta(\alpha = 0.4)$.

The results in terms of combined Sorensen-Dice loss and Surface Dice loss are presented in Table 2. We provide results for HECs as well as individual semantic

Table 1. Detailed specification of 3D convolutional layers (conv) and transposed convolutions (T conv). Every conv layer is followed by normalisation and leaky ReLU activation function.

Network	Layer no	Kernel	Stride	Output dimension	Network	Layer no	Kernel	Stride	Output dimension
conv	0	(1,3,3)	(1,1,1)	(56,256,128)	conv	5	(3,3,3)	(1,1,1)	(7,8,4)
conv	0	(1,3,3)	(1,1,1)	(56,256,128)	conv	5	(3,3,3)	(1,1,1)	(7,8,4)
conv	1	(3,3,3)	(1,2,2)	(56,256,64)	T conv	5	(1,2,2)	(1,2,2)	(7,16,8)
conv	1	(3,3,3)	(1,1,1)	(56,256,64)	conv	4	(3,3,3)	(1,1,1)	(7,16,8)
conv	2	(3,3,3)	(2,2,2)	(28,64,32)	conv	4	(3,3,3)	(1,1,1)	(7,16,8)
conv	2	(3,3,3)	(1,1,1)	(28,64,32)	T conv	4	(2,2,2)	(2,2,2)	(14,32,16)
conv	3	(3,3,3)	(2,2,2)	(14,32,16)	conv	3	(3,3,3)	(1,1,1)	(14,32,16)
conv	3	(3,3,3)	(1,1,1)	(14,32,16)	conv	3	(3,3,3)	(1,1,1)	(14,32,16)
conv	4	(3,3,3)	(2,2,2)	(7,16,8)	T conv	3	(2,2,2)	(2,2,2)	(28,64,32)
conv	4	(3,3,3)	(1,1,1)	(7,16,8)	conv	2	(3,3,3)	(1,1,1)	(28,64,32)
conv	5	(3,3,3)	(1,2,2)	(7,8,4)	conv	2	(3,3,3)	(1,1,1)	(28,64,32)
conv	5	(3,3,3)	(1,1,1)	(7,8,4)	T conv	2	(2,2,2)	(2,2,2)	(56,128,64)
					conv	1	(3,3,3)	(1,1,1)	(56,128,64)
					conv	1	(3,3,3)	(1,1,1)	(56,128,64)
					T conv	1	(1,2,2)	(1,2,2)	(56,256,128)
					conv	0	(3,3,3)	(1,1,1)	(56,256,128)
					conv	0	(3,3,3)	(1,1,1)	(56,256,128)
bottleneck - conv	6	(3,3,3)	(1,2,1)	(7,4,4)					
bottleneck - conv	6	(3,3,3)	(1,1,1)	(7,4,4)					

classes. As can be seen, the CCTNet provides improvement for each of considered hierarchical classes. This is also confirmed by evaluating the performance on individual semantic classes. By closer analysis of the results, we discovered that this improvement comes not only from mixup augmentation alone. We also noticed an improvement after the implementation of the crop in the first stage.

Table 2. Average of Sorensen-Dice loss and Surface Dice loss for predicted hierarchical and semantic classes on KITS21 public data

Network	HECs			Semantic classes		
	Kidney and Masses	Kidney Mass	Tumor	Kidney	Tumor	Cyst
CCTNet (ours)	0.00896	0.00518	0.00236	0.04580	0.01100	0.01026
3D full res nnU-net (baseline)	0.01068	0.00730	0.00302	0.08090	0.02238	0.01804

The Fig. 2 shows the example prediction of the CCTNet. We randomly selected case 0193 from cases containing all three labels (kidney, tumor, cyst). As can be seen, the segmentation of kidneys is pretty good. There is some difference between prediction and ground truth in the cyst located in the right kidney. The most noticeable difference is visible in the tumor label. However, when comparing predictions to the unlabeled CT, we can notice that indeed the real tumor lies outside the ground truth annotated region.

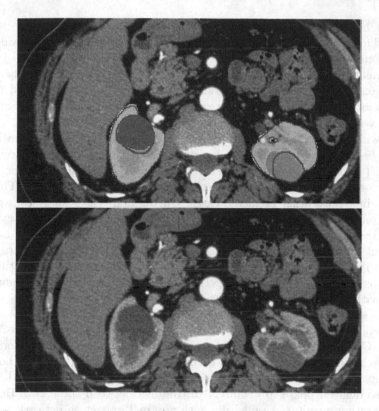

Fig. 2. Example segmentation for the 0193 case. Green contours denote ground truth label. Predicted segmentation classes are: red=kidney, yellow=tumor, blue=cyst. (Color figure online)

Overall performance of our method was also independently measured on KITS21 challenge validation data. Sørensen-Dice and Surface Dice metric was determined for every HEC of every case of the validation set and averaged over each HEC. Additionally average Sørensen-Dice value on the Tumor HEC score was measured and was used as a tiebreaker in case of the same score.

Table 3. Overall performance of CCTNet on KITS21 validation data

Network	Sørensen-Dice	Surface Dice	Tumor Dice
CCTNet	0.8777	0.793	0.795

4 Discussion

We have investigated errors in segmentation of kidney and masses to get a better insight on the network performance. In case of kidney, many errors were at the kidney's boundary. It was interesting to see that in some cases the network provided more accurate boundary than human annotator. However, in evaluation this is considered as a mistake since these two labels do not match.

As was already mentioned, the network tend to mark small cysts where they do not exist. There were also opposite cases, i.e., the network missed to find small cysts. This can be probably expected since these cysts are rather small and the volume is quite heterogeneous, so small cysts are hard to spot.

We will take a closer look at two erroneous predictions depicted in Fig. 3. As can be seen in Fig. 3a, the network marked part of the cyst (brown) as a tumor (yellow). So even though there exists only one single segment (cyst), the network concatenated two different segments (cyst and tumor). This also happens in several other predicted cases. During fault cases analyses, we identified a repeated pattern in true labels, where two distinct segments (cyst and tumor) share common boundaries and therefore create an illusion of one single object (approximately 10–15% of all cysts shared common boundaries with tumor). Therefore, we hypothesize that the network learned this feature and transformed it erroneously to predictions, even though cysts were completely benign with no malign characteristics.

The case depicted in Fig. 3b shows hydronephrosis segmented as a cyst. Hydronephrosis means renal pelvis enlargement, which can resemble kidney cyst on CT scans, when only nephrogenic phase is considered (period when all of the healthy renal parenchyma is contrast-enhanced). In practice, however, hydronephrosis and cysts are two distinct objects and radiologists need to consider also delayed phase to definitely distinguish between those two. However, this was rather a solitary case hardly with any real impact on network learning.

We mentioned just few aspects, but more detailed investigation can reveal more similar cases. Some of these may be corrected by post-processing but for others some updates in network learning would be necessary.

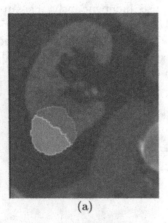 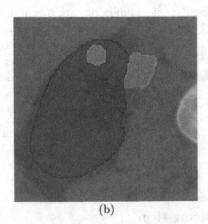

(a)   (b)

Fig. 3. Examples of incorrect segmentation of the proposed network: red represents cysts and yellow tumors. (Color figure online)

Acknowledgment. This work was supported by the Slovak Research and Development Agency under contract No. APVV-16-0211.

References

1. Heller, N., et al.: The state of the art in kidney and kidney tumor segmentation in contrast-enhanced CT imaging: results of the kits19 challenge. Med. Image Anal. **67**, 101821 (2021)
2. Isensee, F., Jaeger, P.F., Kohl, S.A.A., Petersen, J., Maier-Hein, K.H.: nnU-Net: a self-configuring method for deep learning-based biomedical image segmentation. Nat. Meth. **18**(2), 203–211 (2021). https://doi.org/10.1038/s41592-020-01008-z
3. Kotecha, R.R., Motzer, R.J., Voss, M.H.: Towards individualized therapy for metastatic renal cell carcinoma. Nat. Rev. Clinic. Oncol. **16**(10), 621–633 (2019). https://doi.org/10.1038/s41571-019-0209-1
4. Padala, S.A., et al.: Epidemiology of renal cell carcinoma. World J. Oncol. **11**(3), 79–87 (2020). https://pubmed.ncbi.nlm.nlh.gov/32494314
5. Vorontsov, E., Kadoury, S.: Label noise in segmentation networks : mitigation must deal with bias. arXiv:2107.02189 (2021)
6. Zhang, H., Cisse, M., Dauphin, Y.N., Lopez-Paz, D.: Mixup: beyond empirical risk minimization, pp. 1–13. Vancouer, CAN (2018)

Automatic Segmentation in Abdominal CT Imaging for the KiTS21 Challenge

Jimin Heo[✉]

Seoul, Republic of Korea
gjwlals111@gmail.com

Abstract. With the KiTS21 Grand Challenge, I propose the automatic segmentation model between the kidney and the mass of the kidney which includes tumor and cyst. Convolutional Neural Network is trained in patches of three-dimensional abdominal CT imaging. For the segmentation of the 3D image, a variant of U-Net which consists of 3D Encoder-Decoder CNN architecture with additional Skip Connection is used. Lastly, there is a loss function to resolve the class imbalance problem frequently occurring in the task of medical imaging. Sørensen-Dice Score and Surface Dice Score on the test set are 80.13 and 68.61.

Keywords: KiTS21 challenge · 3D Encoder-Decoder U-Net · Medical imaging

1 Introduction

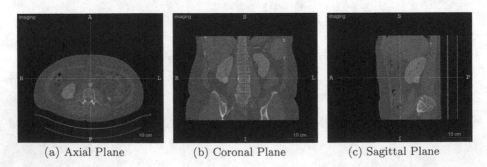

(a) Axial Plane   (b) Coronal Plane   (c) Sagittal Plane

Fig. 1. Examples of KiTS21 challenge dataset. Kidney class is shown in red, tumor class is shown in blue. Not shown in the figure, but cyst class is in green. (Color figure online)

According to Association of American Medical College, a shortage of between 40,800 and 104,900 physicians by 2030 will occur in the United States. Moreover, as reported by International Agency for Research on Cancer, more than 400,000

N. Heller et al. (Eds.): KiTS 2021, LNCS 13168, pp. 98–102, 2022.
https://doi.org/10.1007/978-3-030-98385-7_13

people are affected by kidney cancer in each year and resulted in 175,000 deaths globally. Kidney is involved in removing wastes and excretion of metabolites in the body, and is also responsible for important functions such as moisture and electrolyte balance, acid-alkaline maintenance, and control of other organ functions by producing hormones and vitamins, which may affect surrounding tissues or organs. In addition, tumors in the kidneys are often found after lesions are transmitted to other organs with no special awareness at first, and are often found during tests for other internal diseases due to various non-specific symptoms and signs. The most precise way to evaluate tumors occurring in the kidneys is to take abdominal CT images, so it is important to analyze the acquired images quickly to extract kidneys and tumors. As a way to automate this, the paper presents a method for segmenting abdominal CT images into kidneys, tumors and cysts for KiTS21 which is the grand challenge of Kidney and Tumor Segmentation in 2021 via Deep Convolutional Neural Network model.

2   Methods

Based on U-Net architecture, the winner [3] on KiTS19 developed the variants of the architecture. In the methods, motivated by the winner's model, I propose the model which is the variant of the winner's. With the open source framework MIScnn [5], methods for KiTS21 challenge are implemented in Keras using the Tensorflow backend.

2.1   Training and Validation Data

Our submission made use of the official KiTS21 training set alone. For target data, only voxel-wise majority voting is used.

2.2   Preprocessing

In the 3D Volumetric Image processing, all large dataset cases such as the KiTS dataset should be resampled in common spacing. The reason is that, though the voxel spacings of the cases are generally inconsistent, deep learning neural networks cannot interpret voxel spacings. In order to input the data without that problem, all of the voxel spacings become $3.22 \times 1.62 \times 1.62$ mm. After resampling, because the range of HU (Hounsfield Unit) values is too large to train, I clip the value to $[-79, 304]$ which is the range of fat to soft tissue. Furthermore, Normalization is used following clipping to limit the range and set the standard distribution. Succeeding those preprocessing for the volume, data augmentation is required to regularize overfitting. The method consists of linear and non-linear transformations such as scaling, rotating, symmetric and elastic deformation [7]. Also, Gaussian noise, intensity and contrast algorithms such as gamma correction are included. Lastly, before training, patchwise-crop which analysis of random cropped patches by $80 \times 160 \times 160$ from the image is performed.

2.3 Proposed Method

Although the challenge evaluates only performance, in terms of light-weight and acceleration of both training and inference time, I describe the one-stage semantic segmentation model.

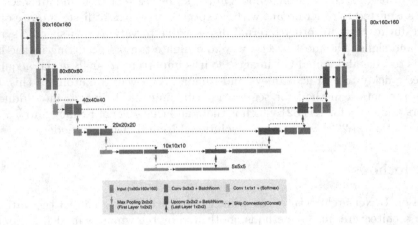

Fig. 2. Network architecture based on 3D Encoder-Decoder U-Net

Network Architecture. It is similar to the winner's model. Basically, both shapes of this model and the winner's model are from 3D U-Net Encoder-Decoder architecture [6]. First, to make a difference and achieve better performance, additional Skip Connections are used. Skip Connection can be split up two ways. One is Residual Connection [1] that is sum of layers. The other is Dense Connection [2] that is concatenation of layers. To maximize the propagation of information, Dense Connection is chosen. Second, Max-pooling is included in the network architecture to extract the maximum value of the feature map. Also, there are Activation Function such as ReLU and Batch Normalization in Transposed Convolution.

Loss Function. In Object Detection or Segmentation, especially in medical tasks such as KiTS21, class imbalance is one of the most important issues. As mentioned, in the challenge, the kidney mass class and even kidney class are much smaller than the background class. To overcome class imbalance, there is Focal loss [4] function which is a variant of Cross-Entropy loss. To use with Softmax, α-factor is modified from scalar to vector. And, there is gamma-factor from the best result in the experiment of Lin et al. [4]. Also, Dice Loss is utilized to perform better. In conclusion, the architecture uses sum of Focal Loss and Dice Loss for loss function.

Optimization and Validation Strategy. For optimization, Adam Optimizer with lr = 3e−4 is chosen. Additionally, monitoring the validation metric on each epoch, a strategy to prevent underfitting is used. The strategy is reducing the learning rate on plateau. If there is no room for improvement in the metric of the validation set, then the scheduler reduces learning rates to induce improvement of the metric. In the paper, the scale factor is 0.1 and patience is 150. The minimum threshold is 1e−4 and the minimum limitation of the learning rate is 3e−6.

3   Results

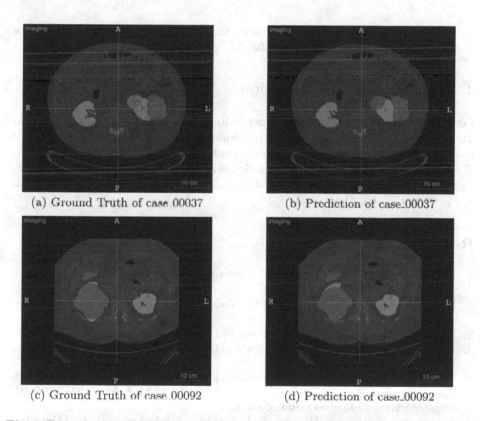

(a) Ground Truth of case_00037   (b) Prediction of case_00037

(c) Ground Truth of case_00092   (d) Prediction of case_00092

Fig. 3. Examples of Ground Truth and Prediction. Kidney class is shown in red, tumor class is shown in blue. cyst class is in green. There is case_00037 predicted incorrectly in the tumor as the cyst. Whereas, case_00092 is predicted more accurately than case_00037. (Color figure online)

Totally, 300 cases are released publicly for the challenge. In the experiment, randomly 240 cases are selected for training and 60 cases are selected for validation.

I train the model on local GPU, NVIDIA RTX 3090 24 GB. Training costs about three days. With batch size of 2 and 500 epochs, The results of the validation score are below.

Table 1. Sørensen-Dice Score and Surface Dice Score on the validation

Kidney Dice	Mass Dice	Tumor Dice	Mean Dice
93.69	78.83	75.03	**82.52**
Kidney SD	Mass SD	Tumor SD	**Mean SD**
87.13	64.05	60.16	**70.45**

On the test, 100 cases are newly used. Sørensen-Dice Score and Surface Dice Score on the test set are **80.13** and **68.61**.

4   Discussion and Conclusion

I described a one-stage semantic segmentation model for KiTS21 Challenge from 3D Abdominal CT imaging. With the model based on U-Net and the sum of Focal Loss and Dice Loss, I attempted to overcome Class Imbalance. As a result, Sørensen-Dice Score and Surface Dice Score on the test set are 80.13 and 68.61. For the better performance, some modules such as Atrous Spatial Pyramid Pooling or U-Net++ could be used.

References

1. He, K., Zhang, X., Ren, S., Sun, J.: Deep residual learning for image recognition (2015)
2. Huang, G., Liu, Z., van der Maaten, L., Weinberger, K.Q.: Densely connected convolutional networks (2018)
3. Isensee, F., Maier-Hein, K.H.: An attempt at beating the 3D U-Net (2019)
4. Lin, T.Y., Goyal, P., Girshick, R., He, K., Dollár, P.: Focal loss for dense object detection (2018)
5. Müller, D., Kramer, F.: MIScnn: a framework for medical image segmentation with convolutional neural networks and deep learning. BMC Med. Imag. **21**(1) (2021). https://doi.org/10.1186/s12880-020-00543-7
6. Ronneberger, O., Fischer, P., Brox, T.: U-Net: convolutional networks for biomedical image segmentation. In: Navab, N., Hornegger, J., Wells, W.M., Frangi, A.F. (eds.) MICCAI 2015. LNCS, vol. 9351, pp. 234–241. Springer, Cham (2015). https://doi.org/10.1007/978-3-319-24574-4_28
7. Simard, P., Steinkraus, D., Platt, J.: Best practices for convolutional neural networks applied to visual document analysis. In: Seventh International Conference on Document Analysis and Recognition. Proceedings, pp. 958–963 (2003). https://doi.org/10.1109/ICDAR.2003.1227801

An Ensemble of 3D U-Net Based Models for Segmentation of Kidney and Masses in CT Scans

Alex Golts[✉], Daniel Khapun, Daniel Shats, Yoel Shoshan, and Flora Gilboa-Solomon

IBM Research, Haifa, Israel
alex.golts@ibm.com

Abstract. Automatic segmentation of renal tumors and surrounding anatomy in computed tomography (CT) scans is a promising tool for assisting radiologists and surgeons in their efforts to study these scans and improve the prospect of treating kidney cancer. We describe our approach, which we used to compete in the 2021 Kidney and Kidney Tumor Segmentation (KiTS21) challenge. Our approach is based on the successful 3D U-Net architecture with our added innovations, including the use of transfer learning, an unsupervised regularized loss, custom postprocessing, and multi-annotator ground truth that mimics the evaluation protocol. Our submission has reached the 2nd place in the KiTS21 challenge.

Keywords: Semantic segmentation · Medical imaging · 3D U-Net · Kidney tumor

1 Introduction

Kidney cancer is among the 10 most frequently diagnosed cancer types [14] and among the 20 deadliest [21]. Surgery is the most common treatment option. Radiologists and surgeons diligently study the appearance of kidney tumors in CT imaging to facilitate optimal treatment prospects [5,13,17]. Automatic segmentation of kidney tumors and surrounding area is a promising tool for assisting them. It has already been proposed as a step in surgery planning [18], as well as enabled medical research relating tumor morphology to surgical outcome [5,13].

The 2019 Kidney and Kidney Tumor Segmentation challenge (KiTS19) [8] was the first to provide a public dataset with kidney tumor labels [9], boosting the available selection of segmentation algorithms specifically designed to segment kidney tumors. In KiTS19, 210 cases were given to participants for training. The kidneys and kidney tumors were annotated and the goal was to segment them accurately in 90 additional test cases.

Compared to KiTS19, the main changes in KiTS21 are:

1. The 90 test cases are now added to the training set which now includes a total of 300 cases. For the 2021 test, 100 additional cases are used.

© Springer Nature Switzerland AG 2022
N. Heller et al. (Eds.): KiTS 2021, LNCS 13168, pp. 103–115, 2022.
https://doi.org/10.1007/978-3-030-98385-7_14

2. A new segmentation class was added to the annotations, denoting cysts. Three Hierarchical Evaluation Classes (HECs) by which participants are evaluated are defined:
    1) **Kidney and Masses:** Kidney + Tumor + Cyst
    2) **Kidney mass:** Tumor + Cyst
    3) **Tumor:** Tumor only
3. A Surface Dice metric [15] was added for evaluation in addition to Sørensen-Dice.
4. Evaluation is performed against a random sample of aggregated segmentation maps that constitute plausible annotations in which different foreground class instances are labeled by different annotators.

Many of the successful algorithms for 3D segmentation in the medical domain are based on 3D variants of the popular U-Net architecture [4,16]. Following its success and dominance as seen in the leading solutions in the KiTS19 challenge [8], we base our solution on the open source nnU-Net framework [12]. It offers automatic configuration of the different stages in a medical imaging segmentation task, including preprocessing, U-Net based network configuration, and optional postprocessing.

Our proposed solution introduces several innovations:

- We employ a label sampling strategy during training to make use of the available multiple annotations and address the new evaluation protocol.
- We perform a form of transfer learning by initializing our network weights with those of a network pretrained on another public medical task.
- We augment the supervised training loss function with an unsupervised regularized term inspired by [7,19] which encourages similar prediction for neighboring voxels with similar intensity.
- We employ postprocessing which removes implausible tumor and cyst predictions that are disconnected from a kidney, as well as small kidney predicted fragments surrounded by another class.

The paper is structured as follows. In Sect. 2 we describe preprocessing and architectural details that were determined automatically by the nnU-Net framework [12]. In Sect. 3 we describe our unique decisions and contribution. These include our annotation sampling method, pretraining, proposed regularized loss, proposed postprocessing algorithm, and choice of models to use in a final ensemble. In Sect. 4 we provide experimental results. Finally, Sect. 5 concludes the paper.

2   nnU-Net Determined Details

2.1   3D U-Net Network Architecture

The U-Net [16] is an encoder-decoder network. The decoder receives semantic information from the end of the encoder (bottom of the "U") and combines it through skip connections with higher resolution features from different layers of the encoder. In our 3D U-Net variant, all convolution kernels are $3 \times 3 \times 3$. Each block in the encoder consists of a sequence of Conv-InstanceNorm [20]-LeakyReLU operations repeated twice. In the encoder, one of these Conv operations has a stride of 2 to facilitate downsampling. In total, there are five downsampling operations. In the decoder, the same number of upsampling operations is done via transposed convolutions.

In most of our experiments, we apply the above architecture as a single-stage network which gets a preprocessed image patch (Sect. 2.3) as input and outputs a final segmentation map. However, we also experimented with a two-stage architecture, described next.

2.2   3D U-Net Cascade Network Architecture

The 3D U-Net cascade is another network type offered in the nnU-Net framework. It serves the purpose of increasing the spatial contextual information that the network sees, while maintaining a feasible input patch size with regards to the GPU memory. This can be achieved by applying a 3D U-Net on downsampled, lower-resolution input data. However, this comes at the cost of reduced detail in the generated segmentations. Therefore, a second stage is performed in which another 3D U-Net is applied on high-resolution input data. In the second stage, the high-resolution input is augmented with extra channels that contain the one-hot encoded segmentation maps generated by the "low-resolution" 3D U-Net from the first stage. These maps are first upsampled to the higher-resolution input data size. Figure 1 depicts the 3D U-Net cascade in high level. In our case, nnU-Net determined the first stage 3D U-Net to be of the same structure as the second stage network, as detailed in Sect. 2.1.

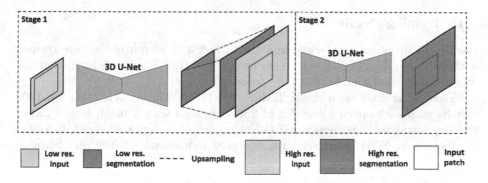

Fig. 1. The 3D U-Net cascade model. In the first stage, low-resolution input enters a 3D U-Net. The patch size covers a large portion of the image, facilitating rich contextual information. Then, the low-resolution segmentation is upsampled and concatenated with the high-resolution input. In the second stage, the input patch covers a smaller portion of the input, but global segmentation information is already available. Then, a second 3D U-Net is applied and refines the segmentations, obtaining them in high resolution.

2.3   Preprocessing

The median voxel spacing in the original training data is $0.78 \times 0.78 \times 3.0$ mm. The median volume shape is $512 \times 512 \times 109$ voxels.

We clip each case's intensity values to the 0.5 and 99.5% of the intensity values in the foreground regions across the training set, which correspond to the range $[-62, 310]$. Then, we subtract the mean and divide by the standard deviation of the intensities in the foreground regions, which correspond to 104.9 and 75.3, respectively.

During training, patches with shape $128 \times 128 \times 128$ are sampled and input to the network. To increase training stability, the patch sampling enforces that more than a third of the samples in a batch contain at least one randomly chosen foreground class.

2.3.1   Low Resolution

In the first stage of the 3D U-Net cascade model, the low-resolution network operates on input data resampled to a common spacing of $1.99 \times 1.99 \times 1.99$ mm. This results in median volume shape of $201 \times 201 \times 207$ voxels for the training cases.

2.3.2   High Resolution

In the single stage 3D U-Net models, and the 2nd stage of the 3D U-Net cascade, the network operates on input data resampled to a common spacing of $0.78 \times 0.78 \times 0.78$ mm. This results in median volume shape of $512 \times 512 \times 528$ voxels for the training cases.

2.4   Training Details

Beside our proposed regularized loss term (Sect. 3.4) all our models are trained with a combination of dice and cross-entropy loss [12]. The loss is applied at the five different resolution levels in the decoder.

Training is done on a single Tesla V100 GPU. The models train for 1000 epochs with each epoch consisting of 250 iterations with a batch size of 2. For the single stage, full-resolution 3D U-Net model, training takes about 48 h.

We use an SGD optimizer with Nesterov momentum of 0.99 and learning rate which decays in each epoch according to $\mathrm{lr} = 0.01 \left(1 - \frac{\text{epoch}}{1000}\right)^{0.9}$.

3   Method

Our solution is based on 3D U-Nets with several novel additions. We found that a 2D U-Net, although faster to train, results in significantly inferior performance on the tumor and cyst classes. For the kidney class, performance is on par with a 3D U-Net. We found at a late stage in the competition that the 3D U-Net cascade model performs better than the single stage 3D U-Net. Therefore, most of our unique contribution was demonstrated using the single stage 3D U-Net model. For the final submission, we ensemble three such models with one cascade model. Next, we describe the training and validation data followed by our unique decisions and contribution, which include the annotation sampling method, optional pretraining, proposed regularized loss, proposed postprocessing algorithm, and choice of models for ensemble.

3.1 Training and Validation Data

We use only the official KiTS21 300 training cases for our submission. The only way in which we *indirectly* use other publicly available data in some of our experiments is by initializing network weights with those of a model pretrained on the Liver Tumor Segmentation (LiTS) database [1].

We train our models on 5 cross-validation splits of 240 training cases and the remaining 60 used for validation. The splits are randomly decided by nnU-Net [12]. In Sect. 4 we report average Dice and Surface Dice scores over the cross-validation splits per HEC, as well as global averages across the HECs. The evaluation metrics are computed using the competition's official evaluation function.

3.2 Pretraining

In Sect. 4.1.2 we show the effect of initializing the network weights from a pretrained model trained on LiTS [1]. This is in contrast to other experiments in which the network weights were initialized randomly. We note that the specific pretrained model we used is available for download under the nnU-Net framework and fits the 3D U-Net network structure determined for our data without any modification. This allowed us a simple way of testing a form of transfer learning for our task.

3.3 Annotations

In KiTS21, each kidney/cyst/tumor instance has been annotated multiple times by different annotators. The competition organizers provided a script for generating (seeded) random plausible aggregated segmentation maps for each case. There are between 6 and 15 such maps per case, depending on the case's number of annotators and class instances. During evaluation, the competition metrics for each case are computed and averaged against all the case's sampled plausible maps.

To resemble the official evaluation protocol, we wanted models to see different plausible annotations during training. We achieve this by choosing for each slice within a case (volume), a random plausible annotation map out of the 6 to 15 options that are available after running the sampling script. We use this random choice as the set of training annotations. In Sect. 4.1.1 we show the effect on performance of using different random seeds for this training annotation selection procedure.

3.4   Regularized Loss

Previous works on weakly supervised segmentation tasks have proposed to add to the "standard" loss term, which makes use of the existing supervised label seeds, another term which is unsupervised. It does not require labels as input, but only the original signal (raw volume in our case), and the network prediction [7,19]. Intuitively, this loss term should encourage the predictions to follow a desired behaviour, such as to be smooth in some sense. In semantic segmentation, we might want the loss to penalize contradicting predictions for neighboring voxels that are similar in their intensity. We experiment with a regularized loss proposed in [7], which can be thought of as a special case of the Potts model [2]. We denote the regularized loss L_{reg}. Then, the total training loss for training our 3D U-Net becomes

$$L_{\text{total}} = L_{\text{dice}} + L_{CE} + \lambda L_{\text{reg}}, \tag{1}$$

where L_{dice} and L_{CE} denote the dice and cross-entropy losses, respectively, and λ is a hyperparameter.

3.4.1   Image Loss

In one of our attempts, we used the regularized loss proposed in [7] for a 2D image, and applied it to each volume slice. This loss affects each pixel through its four upper, lower, right and left neighbors. Let $\mathbf{I}$ be an image; i and j denote two neighbors, ε is the pixel's mentioned 4-neighborhood, and $\mathbf{p}^c$ is the predicted segmentation softmax score for class c. The loss is then given by

$$L_{\text{reg}}(\mathbf{I}, \mathbf{p}) = \sum_c \sum_{(i,j) \in \varepsilon} w_{ij} \left(p_i^c - p_j^c \right)^2, \tag{2}$$

where

$$w_{ij} = e^{-\beta(I_i - I_j)^2}. \tag{3}$$

3.4.2   Volume Loss

Here we generalize the regularized loss to also account for two additional forward and backward neighbors from the adjacent slices. Equations 2–3 remain the same, but ε now contains six neighbors instead of four.

3.5   Postprocessing

We applied a postprocessing algorithm on the segmentation results that removes rarely occurring implausible findings. The algorithm consists of two parts

1. **Tumor and cyst finding positioned outside of kidney findings are removed.**
   We compute a slightly dilated mask of 3D-connected components of voxels classified as tumor or cyst. For each dilated connected component, if it has no intersection with at least one kidney classified voxel, then we change the classification of the corresponding tumor or cyst finding to "background".

2. **Small kidney fragments surrounded by another class are removed.**
We compute 3D-connected components of voxels classified as kidney. We
select all components smaller in volume than the third largest. We then change
the classification of those smaller components to that of the majority of the
voxels in its slightly dilated surrounding.

3.6   Final Submission

For the final submission, we use an ensemble of four models, three single-stage,
high-resolution 3D U-Nets and one 3D U-Net cascade. Each of the four models is
an ensemble on its own of its five trained cross-validation folds. Our postprocessing (Sect. 3.5) was applied on the final segmentation output after the ensemble.
The following is a description of each model in our final ensemble:

1. **3D U-Net** trained with the regularized loss from Sect. 3.4.1.
2. **3D U-Net** for which training was initialized with a model pretrained on
   LiTS. (Sect. 3.2).
3. **3D U-Net** trained with a different random seed for the training annotation
   generation process (Sect. 3.3) than the other three models in the ensemble.
4. **3D U-Net cascade** in which training of the first-stage, low-resolution network was initialized with a model pretrained on LiTS (Sect. 3.2).

4   Results

In all experiments we use the official evaluation code which calculates Dice and
Surface Dice metrics averaged across sampled plausible annotation maps. The
results we show here are all average scores across five cross-validation splits.
For brevity we denote in the tables in this section the per-HEC dice scores as
D1, D2, D3 for the "Kidney and Masses", "Kidney mass" and "Tumor" HECs,
respectively, and their mean is denoted MD. Similarly, surface dice scores are
denoted SD1, SD2, SD3, and their mean MSD.

4.1   Single-Stage, High-Resolution 3D U-Net

The following experiments were made based on the single-stage 3D U-Net model
(Sect. 2.1).

4.1.1   Random Annotations

In Table 1 we show how our 3D U-Net trained with random annotation procedure
as described in Sect. 3.3 performed against a baseline that we trained which used
aggregated maps according to a majority vote (MAJ). We also show results of
the baseline published in [11].

We see that the random annotation procedure unfortunately has no significant effect on performance. We still decided to add the model with seed 3 to our
final model ensemble. In all the next experiments, we use our random annotation
procedure with seed 1.

Table 1. Our random annotation procedure vs. baseline MAJ annotations

Model	D1	D2	D3	SD1	SD2	SD3	MD	MSD
Baseline [11]	0.9666	0.8618	0.8493	0.9336	0.7532	0.7371	0.8926	0.8080
Baseline (our)	0.9660	0.8589	0.8444	0.9334	0.7506	0.7320	0.8897	0.8053
Seed 1	0.9662	0.8583	0.8449	0.9335	0.7513	0.7330	0.8898	0.8059
Seed 2	0.9655	0.8581	0.8419	0.9324	0.7500	0.7297	0.8885	0.8040
Seed 3	0.9668	0.8645	0.8478	0.9347	0.7567	0.7356	**0.8930**	**0.8090**

4.1.2 Pretraining

In Table 2, we show the effect of transfer learning, namely initializing our 3D U-Net model from weights of a model pretrained on LiTS.

Table 2. Transfer learning from LiTS

Model	D1	D2	D3	SD1	SD2	SD3	MD	MSD
Baseline [11]	0.9666	0.8618	0.8493	0.9336	0.7532	0.7371	0.8926	0.8080
Baseline (our)	0.9660	0.8589	0.8444	0.9334	0.7506	0.7320	0.8897	0.8053
With pretraining	0.9674	0.8651	0.8518	0.9346	0.7563	0.7396	**0.8948**	**0.8102**

We see some improvement, therefore we add the model with initialization from a model pretrained on LiTS to our final model ensemble.

4.1.3 Regularized Loss

In Table 3 we show the effect of adding regularized loss, as described in Sect. 3.4. Following limited hyperparameter search, we use $\beta = 10$ for the image loss, $\beta = 5$ for the volume loss, and $\lambda = 1$ for both versions. We add the model with image regularized loss to our final model ensemble, as it showed slight improvement over at least *our* baseline. Our experiments showed that hyperparameter tuning for β and λ are important for the regularized loss. This could be one of the reasons we did not manage to better optimize the volumetric version of the loss within our time and resources constraints. It is also why we did not choose to employ

Table 3. Performance with regularized loss

Model	D1	D2	D3	SD1	SD2	SD3	MD	MSD
Baseline [11]	0.9666	0.8618	0.8493	0.9336	0.7532	0.7371	**0.8926**	**0.8080**
Baseline (our)	0.9660	0.8589	0.8444	0.9334	0.7506	0.7320	0.8897	0.8053
Image loss	0.9659	0.8615	0.8493	0.9341	0.7528	0.7370	0.8922	**0.8080**
Volume loss	0.9663	0.8609	0.8482	0.9336	0.7512	0.7337	0.8918	0.8062

this loss in conjunction with other steps like initializing with a pretrained model, or the next experiment with the 3D U-Net cascade model. We suspected that separate hyperparameter tuning might need to be performed for each scenario.

4.2 3D U-Net Cascade

In Table 4 we show the performance of our trained 3D U-Net cascade model (Sect. 2.2) compared to the baseline published in [11] (for the same model type).

Table 4. Performance of our 3D U-Net cascade model

Model	D1	D2	D3	SD1	SD2	SD3	MD	MSD
Cascade baseline [11]	0.9747	0.8799	0.8491	0.9453	0.7714	0.7393	0.9012	0.8187
Our cascade	0.9747	0.8810	0.8583	0.9459	0.7709	0.7461	**0.9046**	**0.8210**

We see some improvement for our cascade model over the published baseline. This could be due to our random annotation procedure (Sect. 3.3) and our initialization of the first-stage, low-resolution 3D U-Net with a model pretrained on LiTS (Sect. 3.2).

4.3 Model Ensemble

In Table 5 we show the effect of ensembling four models as described in Sect. 3.6. We also show for comparison an ensemble of only the three single-stage 3D U-Net models, as well as the best single models (without ensemble), out of both single-stage and cascade. Again, metrics are averaged across all five cross-validation splits.

Table 5. Model ensemble

Model	D1	D2	D3	SD1	SD2	SD3	MD	MSD
Best model (cascade)	0.9747	0.8810	0.8583	0.9459	0.7709	0.7461	**0.9046**	0.8210
Best 1-stage model	0.9674	0.8651	0.8518	0.9346	0.7563	0.7396	0.8948	0.8102
Ensemble of 1-stage models	0.9674	0.8667	0.8535	0.9363	0.7610	0.7442	0.8959	0.8138
Final ensemble	0.9702	0.8751	0.8597	0.9400	0.7709	0.7525	0.9017	**0.8211**

We see that the cascade model, which outperformed all the others, is alone better than the final ensemble in terms of the average Dice score. But we do see a slight improvement in average Surface Dice, and also in the tumor class metrics, which are arguably the most critical in practice (and also the tumor Dice is used as a tiebreaker in KiTS21). We also see that ensembling the three single-stage 3D U-Net models improves over the best model among them. Therefore we decided to ensemble all four models in our final submission.

112   A. Golts et al.

4.4 Postprocessing

In Table 6 we show the result of applying our proposed postprocessing algorithm
(Sect. 3.5) to the segmentation results of our final ensemble. Again, metrics are
averaged across all five cross-validation splits. We see improvement across all
metrics.

Table 6. Results with postprocessing applied to our final ensemble segmentations

Model	D1	D2	D3	SD1	SD2	SD3	MD	MSD
Without postprocessing	0.9702	0.8751	0.8597	0.9400	0.7709	0.7525	0.9017	0.8211
With postprocessing	0.9715	0.8790	0.8638	0.9415	0.7751	0.7569	**0.9047**	**0.8245**

Table 7 shows an example from case 16 in the database, predicted using our
model ensemble. Specifically, for the demonstration to be fair, the ensembled
model of cross validation fold 0, in which case 16 was part of the validation set.
In the first row, we see slice 60, which contains kidney (red) and tumor (green)
findings. In the second row, we see slice 105, in which a false tumor finding was
predicted, and successfully removed after applying our postprocessing algorithm,
since it has no contact with a kidney prediction.

Table 7. Example predictions for case 16. Top row: Slice 60, which contains kidney
(red) and tumor (green) findings. Bottom row: Slice 105, which contains a false tumor
prediction, successfully removed by our postprocessing algorithm.

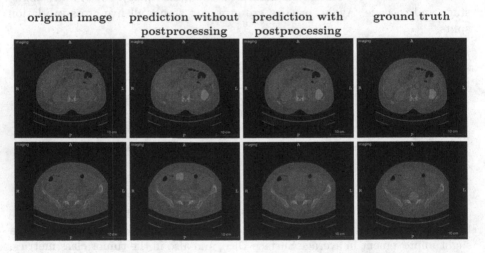

4.5   Test Set Results

In Table 8 we show the results of the top-five submissions on the KiTS21 test set, where our submission reached second place.

Table 8. KiTS21 test set results

Rank	Team	D1	D2	D3	SD1	SD2	SD3	MD	MSD
1	Z. Zhao et al. [3]	0.977	0.886	0.860	0.957	0.772	0.749	0.908	0.826
2	**A. Golts et al.**	0.976	0.881	0.832	0.956	0.769	0.722	0.896	0.816
3	Y. George [6]	0.976	0.876	0.831	0.955	0.765	0.722	0.894	0.814
4	X. Yang et al. [23]	0.973	0.874	0.822	0.950	0.758	0.707	0.890	0.805
5	M. Wu and Z. Liu [22]	0.970	0.863	0.811	0.944	0.747	0.711	0.881	0.801

5   Discussion and Conclusions

We presented results of our 3D U-Net based approach to solving the KiTS21 challenge. We managed to demonstrate minor improvements over published baselines based on a single-stage 3D U-Net, as well as a two-stage 3D U-Net cascade. Improvements are owed, to varying degrees, to our following contributions: a method for utilizing multiple annotations during training, weight initialization from a model pretrained on a different task, an unsupervised term added to the loss function that encourages smoothness in the segmentation predictions, ensembling of multiple models, and a proposed postprocessing algorithm.

The participation in the challenge leaves us with quite a few interesting directions for future research. We experimented with using image blending techniques for injecting tumor volume regions onto healthy kidney regions in order to enrich the training dataset. These efforts showed initial promise but did not materialize in time for the competition deadline. We believe this could be a promising direction for future research. We now realize that better performance could be reached if we applied some of our contributions to the better performing 3D U-Net cascade model, rather than the single-stage 3D U-Net. The regularized loss could benefit from more thorough hyperparameter tuning as well as further generalization to use more neighboring voxels. Additionally, more recent network architectures for semantic segmentation are worth exploring. Outside the scope of this particular challenge, it is worth investigating the trade-off between accuracy and runtime in medical imaging segmentation, for example as is evident when comparing 2D and 3D U-Net architectures.

As we look to experiment with different network architectures, or work on extending the idea of regularized losses for medical imaging segmentation, we may opt for open source frameworks designed for flexible and efficient research in the medical imaging domain. One such promising framework is the recently released FuseMedML library [10].

Acknowledgments. We thank the KiTS competition organizers, data providers, and annotators for their great effort in advancing the science around kidney cancer, including improving algorithms for automatic segmentation. We further thank the creator and contributors to the nnU-Net framework, an excellent and generally applicable baseline for medical segmentation tasks.

References

1. Bilic, P., et al.: The liver tumor segmentation benchmark (LiTS). arXiv preprint arXiv:1901.04056 (2019)
2. Boykov, Y., Veksler, O., Zabih, R.: Fast approximate energy minimization via graph cuts. IEEE Trans. Pattern Anal. Mach. Intell. **23**(11), 1222–1239 (2001)
3. Chen, Z.Z.: A coarse-to-fine framework for the 2021 kidney and kidney tumor segmentation challenge (2021). https://openreview.net/forum?id=6Py5BNBKoJt
4. Çiçek, Ö., Abdulkadir, A., Lienkamp, S.S., Brox, T., Ronneberger, O.: 3D U-Net: learning dense volumetric segmentation from sparse annotation. In: Ourselin, S., Joskowicz, L., Sabuncu, M.R., Unal, G., Wells, W. (eds.) MICCAI 2016. LNCS, vol. 9901, pp. 424–432. Springer, Cham (2016). https://doi.org/10.1007/978-3-319-46723-8_49
5. Ficarra, V., et al.: Preoperative aspects and dimensions used for an anatomical (PADUA) classification of renal tumours in patients who are candidates for nephron-sparing surgery. Eur. Urol. **56**(5), 786–793 (2009)
6. George, Y.M.: A coarse-to-fine 3D U-Net network for semantic segmentation of kidney CT scans (2021). https://openreview.net/forum?id=dvZiPuZk-Bc
7. Golts, A., Freedman, D., Elad, M.: Deep energy: task driven training of deep neural networks. IEEE J. Sel. Top. Sig. Process. **15**(2), 324–338 (2021)
8. Heller, N., et al.: The state of the art in kidney and kidney tumor segmentation in contrast-enhanced CT imaging: results of the KiTS19 challenge. Med. Image Anal. **67**, 101821 (2021)
9. Heller, N., et al.: The KiTS19 challenge data: 300 kidney tumor cases with clinical context, CT semantic segmentations, and surgical outcomes. arXiv preprint arXiv:1904.00445 (2019)
10. IBM Research, Haifa: FuseMedML (2021). https://doi.org/10.5281/ZENODO.5146491. https://zenodo.org/record/5146491. https://github.com/IBM/fuse-med-ml
11. Isensee, F.: nnU-Net baseline for the KiTS21 task (2021). https://github.com/neheller/kits21/tree/master/examples/nnUNet_baseline
12. Isensee, F., Jaeger, P.F., Kohl, S.A., Petersen, J., Maier-Hein, K.H.: nnU-Net: a self-configuring method for deep learning-based biomedical image segmentation. Nat. Methods **18**(2), 203–211 (2021)
13. Kutikov, A., Uzzo, R.G.: The RENAL nephrometry score: a comprehensive standardized system for quantitating renal tumor size, location and depth. J. Urol. **182**(3), 844–853 (2009)
14. National Cancer Institute: Common cancer types (2021). https://www.cancer.gov/types/common-cancers
15. Nikolov, S., et al.: Deep learning to achieve clinically applicable segmentation of head and neck anatomy for radiotherapy. arXiv preprint arXiv:1809.04430 (2018)

16. Ronneberger, O., Fischer, P., Brox, T.: U-Net: convolutional networks for biomedical image segmentation. In: Navab, N., Hornegger, J., Wells, W.M., Frangi, A.F. (eds.) MICCAI 2015. LNCS, vol. 9351, pp. 234–241. Springer, Cham (2015). https://doi.org/10.1007/978-3-319-24574-4_28
17. Simmons, M.N., Ching, C.B., Samplaski, M.K., Park, C.H., Gill, I.S.: Kidney tumor location measurement using the C index method. J. Urol. **5**, 1708–1713 (2010)
18. Taha, A., Lo, P., Li, J., Zhao, T.: Kid-Net: convolution networks for kidney vessels segmentation from CT-volumes. In: Frangi, A.F., Schnabel, J.A., Davatzikos, C., Alberola-López, C., Fichtinger, G. (eds.) MICCAI 2018. LNCS, vol. 11073, pp. 463–471. Springer, Cham (2018). https://doi.org/10.1007/978-3-030-00937-3_53
19. Tang, M., Perazzi, F., Djelouah, A., Ayed, I.B., Schroers, C., Boykov, Y.: On regularized losses for weakly-supervised CNN segmentation. In: Ferrari, V., Hebert, M., Sminchisescu, C., Weiss, Y. (eds.) ECCV 2018. LNCS, vol. 11220, pp. 524–540. Springer, Cham (2018). https://doi.org/10.1007/978-3-030-01270-0_31
20. Ulyanov, D., Vedaldi, A., Lempitsky, V.: Instance normalization: the missing ingredient for fast stylization. arXiv preprint arXiv:1607.08022 (2016)
21. Wikipedia: list of cancer mortality rates in the united states (2021). https://en.wikipedia.org/wiki/List_of_cancer_mortality_rates_in_the_United_States
22. Wu, M., Liu, Z.: Less is more (2021). https://openreview.net/forum?id=immB02xhM15
23. Yang, X., Jianpeng, Z., Yong, X.: Transfer learning for KiTS21 challenge (2021). https://openreview.net/forum?id=XXtHQy0d8Y

Contrast-Enhanced CT Renal Tumor Segmentation

Chuda Xiao[1], Haseeb Hassan[1,2], and Bingding Huang[1(✉)]

[1] College of Big Data and Internet, Shenzhen Technology University, Shenzhen, China
huangbingding@sztu.edu.cn
[2] Guangdong Key Laboratory for Biomedical Measurements and Ultrasound Imaging, School of Biomedical Engineering, Shenzhen University Health Science Center, Shenzhen, China

Abstract. Automated detection and segmentation of kidneys, tumors, and cysts are useful for renal diagnosis and treatment planning. Here we propose a two-stage contrast-enhanced CT detection and segmentation framework that automatically segments the kidney, kidney tumor, and cyst. Testing the proposed algorithm on the KiTS21 dataset, we achieve the mean dice of 0.5905 and the mean surface dice of 0.4234.

Keywords: Two-stage network · Kidney tumor · Segmentation

1 Introduction

Renal refers to the kidneys. The terms "tumor" and "mass" refer to abnormal body growths. Our kidneys might develop masses (growths or tumors) from time to time. Some kidney tumors are benign (noncancerous), whereas others are malignant (cancerous). According to GLOBOCAN data from 2018, an estimated 403,000 persons are diagnosed with abnormal kidney growth each year, accounting for 2.2% of all cancer diagnoses [1]. Due to the enormous variation in kidney and kidney tumor shape, there is a lot of interest in understanding how tumor morphology influences surgical outcomes [2, 3] and developing sophisticated surgical planning techniques [4].

For this purpose, measuring the shape and dimensions of a kidney tumor can be revealed by contrast-enhanced Computed Tomography (CT) imaging which is essential for diagnosis, treatment, and safe surgery [5]. Safe surgery involves avoiding injury to the kidney's vascular network. As a result, automatic semantic segmentation becomes a critical component of surgical planning and is widely used. Previously the KiTS2019 [6] focused on kidney and kidney tumor segmentation, whereas the newly introduced challenge KiTS21 includes an additional class of cyst. Therefore, in this article, we propose to segment kidney, kidney tumor, and cyst. This remaining manuscript is organized as follows. Methods and procedures are defined in Sect. 2, Sect. 3 presents the experimental results, and Sect. 4 concludes the overall manuscript.

© Springer Nature Switzerland AG 2022
N. Heller et al. (Eds.): KiTS 2021, LNCS 13168, pp. 116–122, 2022.
https://doi.org/10.1007/978-3-030-98385-7_15

2   Methods

In this work, we use a multi-stage algorithm to segment kidney, tumor and cyst. The proposed algorithm comprises two stages, as depicted in Fig. 1. The first stage involves the detection process, and the second stage applies the segmentation process. In both settings, we consider ResUnet 3D as the backbone network. The detection process we use for accurate localization of the kidney. The reason for this, to make the subsequent segmentation process more effective.

Primarily, we preprocess the CT training data by resampling (the z-axis spacing to 2 mm, while retaining the x, y-axis unchanged) and cropping them to sizes $32 \times 384 \times 384$. These CT images are provided as an input to the detection network. The detection network initially detects both the kidneys' shapes based on the preprocessed inputs. After the detection process, we calculate the centers of each detected kidney according to (x, y, z) points by using the *skimage* library. Further, the detected kidneys were cropped into cube (volume) sizes $64 \times 128 \times 128$ and provided that to the segmentation network to predict kidney, tumor, and cyst. After that, we combine all the predicted masks from the segmentation network to the volume size ($s \times 384 \times 384$). Finally, we post-process (by padding and resampling) the combined volume size ($s \times 384 \times 384$) to the original CT scan size, i.e. ($n \times 512 \times 512$).

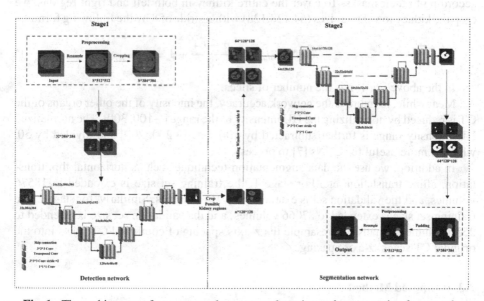

Fig. 1. The architecture of our proposed two-stage detection and segmentation framework.

2.1   Training and Validation Data

To validate our proposed framework, we use the official KiTS21 dataset. The dataset is divided into training and validation sets (training: 269 cases, validation: 30 cases). For instance, 00000 to 00269 are training cases, and 00270 to 00299 are validation cases. Note that we removed case 00160 from the adopted dataset due to its different size. In addition, we use voxel-wise majority voting (MAJ) for training and validation.

2.2 Preprocessing

After analyzing the KiTS21 data, we found different z-axis spacing between cases, i.e., in the range [0.5 mm–5 mm]. In the dataset, around 125 cases where the z-axis spacing is 5mm. To balance the z-axis spacing between each case, we resample the data to size $(s, 512, 512)$ and set the z-axis spacing to 2mm, while the x-axis and y-axis remain unchanged.

Further, we discard the useless regions from each resampled CT slice and crop it to size $(s, 384, 384)$ according to the image's center point for both training and prediction. Note that s denotes the number of resampled slices. The reasons for cropping are to reduce the image size and to cover the possible kidney regions. To improve the GPU utilization, we use the sliding window with a stride of 8 to resize the CT images to the volume size (32, 384, 384). We also perform some data cleaning by removing those volumes not containing the kidney, tumor, and cyst.

After the kidneys detection in stage 1, we separate left and right kidney regions according to their masks. To cover the entire kidney in both left and right regions, we crop the kidney to volume sizes (64,128,128) with a stride of 8 by the following formula.

$$Counter_{volume} = \frac{n - 64}{8} + 1 \tag{1}$$

In the above Eq. 1, n is the number of slices.

Meanwhile, to improve the network accuracy, the intensity of the other organs of the CT is reduced by normalizing the HU intensity to the range [−100, 300]. The normalized HU intensity range is further subtracted by 100, i.e., [−200, 200] and divided by 50, which is more useful for CNNs [7] to process.

In addition, we use the data augmentation technique such as horizontal flip, translation, affine translation, etc. For stage 1, the training set size is extended to 18576 volumes, and the validation set is extended to 1914 volumes. Similarly, for stage 2, the training set size is extended to 27066 volumes, and the validation set size is extended to 2550 volumes. Finally, we resample the z-axis spacing of combined CT scans into the original CT scans' z-axis spacing.

2.3 Proposed Method

2.3.1 Network Architecture

Since U-Net [8] has achieved excellent segmentation results specifically for 3D volumetric CT scans, which are 2D image sequences, therefore, our intended model also takes advantage of U-Net 3D [1] and the Residual network [9] to perform the three-class segmentation task. Our proposed ResU-Net 3D network architecture is shown in Fig. 2.

The intended network uses three encoder and decoder blocks, which are residual convolutional blocks. The residual block contains three convolutional layers, and the

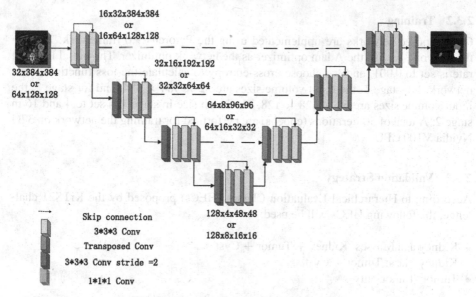

16x32x384x384
or
16x64x128x128

32x384x384
or
64x128x128

32x16x192x192
or
32x32x64x64

64x8x96x96
or
64x16x32x32

128x4x48x48
or
128x8x16x16

Skip connection
3*3*3 Conv
Transposed Conv
3*3*3 Conv stride =2
1*1*1 Conv

Fig. 2. The proposed ResU-Net3D network

residual convolution kernel size is $3 \times 3 \times 3$. The stride size of every residual convolution is $1 \times 1 \times 1$. In addition, the up-sampling using the transposed convolution and the up-sampling's convolution kernel size is $2 \times 2 \times 2$ where the stride is 2. The down-sampling stride size is $2 \times 2 \times 2$ convolution, and the final output layer is $1 \times 1 \times 1$ convolution. The network parameters are given in Table 1.

Table 1. The network parameters

Name	Layers	Stride	Kernel size	Padding
Convolution (C)	Conv	–	–	
	Batch norm	–	–	
	ReLU	–	–	
Residual block	C1	$1 \times 1 \times 1$	$3 \times 3 \times 3$	$1 \times 1 \times 1$
	C2	$1 \times 1 \times 1$	$3 \times 3 \times 3$	$1 \times 1 \times 1$
	C3	$1 \times 1 \times 1$	$1 \times 1 \times 1$	0
	C1 + C3			
Down sample	C	$2 \times 2 \times 2$	$2 \times 2 \times 2$	0
Up sample	Transpose Conv	$2 \times 2 \times 2$	$2 \times 2 \times 2$	0
Concat	Residual Block + Up Sample			
Output layer	C	$1 \times 1 \times 1$	$1 \times 1 \times 1$	0

2.3.2 Training

Our proposed networks are implemented using the Pytorch1.9.0 framework. To train the network, we use the Adam optimizer as the network optimizer. The initial learning rate is set to 0.001, and we choose cross-entropy to calculate the loss function of the network. For stage 1, the input volume sizes are $32 \times 384 \times 384$, and for stage 2, the input volume sizes are $64 \times 128 \times 128$. The batch size in stage 1 is set to 4 and 16 in stage 2. A total of 50 iterations (epochs) are performed for training the network on 32G Nvidia V100 GPU.

2.3.3 Validation Strategy

According to Hierarchical Evaluation Classes (HECs) proposed by the KiTS21 challenge, the following HECs will be used.

- Kidney and Masses: Kidney + Tumor + Cyst
- Kidney Mass: Tumor + Cyst
- Tumor: Tumor only

To evaluate the performance of the model, we use dice and surface dice(SD).

3 Results

We selected case00270~case00299 as the validation set and provided these cases' to the proposed pipeline for prediction purposes. Table 2 shows the three classes dice and surface dice where KMC denotes kidney and Masses class, Kidney Mass is denoted by KM, and Tumor class is denoted by T. Table 3 shows the achieved Mean Dice and surface dice are 0.5905 and 0.4234 in test set. Figure 3 provides the visual analysis of the proposed pipeline prediction.

Table 2. Mean Dice and SDS of classes on KiTS21 validation set

Network	KMC Dice	KM Dice	T Dice	KMC SD	KM SD	T SD
Ours	0.9130	0.5635	0.4864	0.7718	0.3424	0.2834

Table 3. Mean Dice and SDS of the proposed pipeline on KiTS21 test set

Network	Mean Dice	Mean SD
Ours	0. 5905	0. 4234

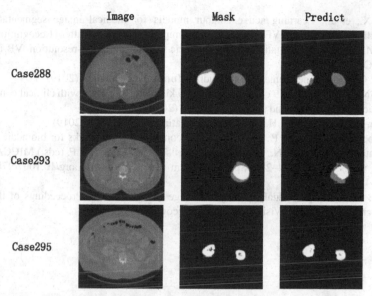

Fig. 3. Visualization of predictions of our proposed model. The 1st column is the input CT slice, the 2nd column is the mask, and the third column is our proposed method predictions.

4 Discussion and Conclusion

This work proposed two-stages detection and segmentation architecture to automatically segment kidney, cyst, and tumor based on the KiTS21 benchmark. For both the detection and segmentation networks, the ResUnet 3D is utilized as the backbone. The designed two-stage architecture achieved the mean dice of kidney and messes, kidney messes and the tumor is 0.5905, and the mean surface dice is 0.4234. However, our model generated low dice and surface dice for the tumor and cyst. The reason for that is that tumor and cyst are quite tiny and have limited availability in the adopted dataset. Therefore, to address this problem, in the future, we plan to augment data of the tumor and cyst for better detection and segmentation, which will eventually lead us to better quantitative outcomes.

Acknowledgment. We would like to express our gratitude to the KITS21 organizers and the Shenzhen Technology University School-Enterprise Graduate Student Cooperation Fund.

References

1. Global cancer observatory: cancer today. International Agency for Research on Cancer, Lyon, France. https://gco.iarc.fr/today. Accessed 2 Mar 2020
2. Çiçek, Ö., Abdulkadir, A., Lienkamp, S.S., Brox, T., Ronneberger, O.: 3D U-Net: learning dense volumetric segmentation from sparse annotation. In: Ourselin, S., Joskowicz, L., Sabuncu, M.R., Unal, G., Wells, W. (eds.) MICCAI 2016. LNCS, vol. 9901, pp. 424–432. Springer, Cham (2016). https://doi.org/10.1007/978-3-319-46723-8_49

3. Chen, X., et al.: Learning active contour models for medical image segmentation. In: Proceedings of the IEEE/CVF Conference on Computer Vision and Pattern Recognition (2019)
4. Han, M., et al.: Segmentation of CT thoracic organs by multi-resolution VB-nets. In: SegTHOR@ ISBI (2019)
5. Liu, S.: Coarse to Fine Framework for Kidney Tumor Segmentation (2019)
6. Heller, N., et al.: The kits19 challenge data: 300 kidney tumor cases with clinical context. CT semantic segmentations, and surgical outcomes (2019)
7. Isensee, F., Maier-Hein, K.H.: An attempt at beating the 3D U-Net (2019)
8. Ronneberger, O., Fischer, P., Brox, T.: U-net: convolutional networks for biomedical image segmentation. In: Navab, N., Hornegger, J., Wells, W.M., Frangi, A.F. (eds.) MICCAI 2015. LNCS, vol. 9351, pp. 234–241. Springer, Cham (2015). https://doi.org/10.1007/978-3-319-24574-4_28
9. He, K., et al.: Deep residual learning for image recognition. In: Proceedings of the IEEE Conference on Computer Vision and Pattern Recognition (2016)

A Cascaded 3D Segmentation Model for Renal Enhanced CT Images

Dan Li[1,2], Zhuo Chen[2], Haseeb Hassan[1,3], Weiguo Xie[2], and Bingding Huang[1(✉)]

[1] College of Big Data and Internet, Shenzhen Technology University, Shenzhen, China
huangbingding@sztu.edu.cn
[2] Wuerzburg Dynamics Inc, Shenzhen, China
[3] Guangdong Key Laboratory for Biomedical Measurements and Ultrasound Imaging, School of Biomedical Engineering, Shenzhen University Health Science Center, Shenzhen, China

Abstract. In order to compete in the KiTS21 challenge, we propose a 3D deep learning cascaded model for the renal enhanced CT image segmentation. The proposed model comprises two stages, where stage 1 segments the kidney and stage 2 segments the tumor and cyst. The proposed deep learning network architecture is based on the residual and 3D UNet architecture. The designed network is utilized for each segmentation stage (for stage 1 and stage 2). Our intended cascaded model achieved a dice score of 0.96 for the kidney, 0.81 for the tumor, and 0.45 for the cyst on the KiTS21 validation dataset.

Keywords: Renal segmentation · Renal tumor segmentation · Renal cyst segmentation

1 Introduction

Every year around 400,000 people are affected by kidney tumors. Due to the wide variety in kidney and kidney tumor morphology, there is currently a great interest in tumor morphology and its surgical outcomes [1]. For instance, focusing on kidney morphology is essential to advance surgical planning. In order to accelerate such research and development, KiTS challenge has been introduced. In KiTS 2019, the initial focus was on kidney and kidney tumor segmentation [2]. However, the ongoing KiTS 21 challenge focuses on three-class segmentation tasks: kidney, kidney tumor, and cyst segmentation. Automatic semantic segmentation is a promising tool for these endeavors, but morphological variability is not easy. Therefore, there is a need for reliable segmentation techniques that can perform well and provide aid to renal surgical planning. Here we propose a cascaded 3D model that efficiently performs the renal, tumor, and cyst segmentation to solve the challenging task of morphological variability.

D. Li and Z. Chen—These authors contribute equally to this work.

2 Methods

Our intended segmentation approach consists of a two-stage segmentation model based on 3D-Unet. For this purpose, we adopt different pre-and post-processing strategies. First, we resample the original CT scans and then crop the center parts of the CT scans. The cropped CT scans are then used as the input to the segmentation network, which is the first stage of our proposed model for renal contour prediction.

Further, a histogram equalization [3] approach is applied on segmented/predicted renal to enhance the contained tumor and cyst. Because the histogram equalization uplift the pixel brightness (contrast) and useful for the subsequent segmenting step [4]. Afterward, we concatenate the segmented renal (from stage 1) and enhanced renal (after the enhancement process) as two channels and provided that as input to stage 2 (second segmentation network). Upon providing the concatenated two channels input to the stage 2 network, the intended model predicts the tumor and cyst. Finally, the prediction results of both stages are merged into a single channel and reduced to the original CT size to evaluate the model performance. The algorithmic flow of the proposed pipeline is depicted in Fig. 1.

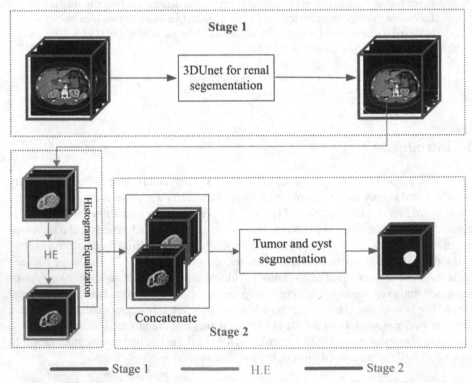

Fig. 1. Our proposed cascaded 3D model involves two stages. Stage 1 segments the renal, and stage 2 performs tumor and cyst segmentation. Finally, all the predicted outputs from each stage are merged into a single channel output.

2.1 Training and Validation Data

To train our model, we adopt the KiTS21 benchmark dataset. The benchmark dataset contains 300 contrast-enhanced CT scans, which provide three-class labels for kidney, kidney tumor, and cyst. We divide the KiTS21 dataset into training and validation sets, i.e., 270 and 30, respectively. In addition, we use voxel-wise majority voting (MAJ) for training and validation.

2.2 Preprocessing

As an initial preprocessing step, the CT intensities (Hounsfield units-HU) of training and validation sets were selected to a range of [-135, 215]. Doing so will change the appearance of the picture to highlight particular structures [8]. Further, we analyzed the spacing distribution of KiTS 21 CT images. After analyzing them, we found different z-axis spacing between cases, i.e., in the range [0.5mm-5mm]. To balance the z-axis spacing between each case, we resample the data and set the z-axis spacing to 3.0mm. We also use min-max normalization to normalize the HU intensities. For kidney segmentation stage 1, we resize the x and y-axis of the input data to (256, 256). For stage 2 (tumor and cyst segmentation), the respective regions of the preprocessed image (512, 512) is cropped according to the predicted labels (masks) from stage 1.

2.3 Proposed Method

2.3.1 Architecture

Our proposed network uses the Residual 3D U-Net [5–7], which has four up-sampling layers and four down-sampling layers. Each layer is composed of 3D convolution, ReLU activations, and batch normalization. The first level of the UNet extracts 32 feature maps in the proposed pipeline, and each down-sampling process maximizes the extracted feature maps up to 512. The network learning rate is 0.001; the batch size is 8, epochs are 200, and cross-entropy is used as the loss function. We utilize the Adam optimizer as the network optimizer to train the network. The proposed architecture is used for each stage separately. The Pytorch1.9.0 framework is used to implement our proposed approach (Fig. 2).

2.3.2 Methodology

Stage 1: Kidney segmentation

In the kidney segmentation stage, we initially aim to extract the whole kidney. Because the dataset contains multiple classes such as kidney, tumor, and cyst, thus it is not easy to directly detect or segment the kidney. Therefore, we change the individual kidney, tumor, and cyst masks into a single class, i.e., kidney. Considering the z-axis, we crop the 3D cube with length 96 and stride 48 by using the center crop on the x and y-axis to clip the subsequent cube with shape (1,96,192,192) and provide that as input to the kidney segmentation network. For this purpose, we use the regionprops function to analyze all input data's kidney regions and try different shape sizes, where we find the shape size (1,96,192,192) is a perfect range to cover both kidney regions entirely.

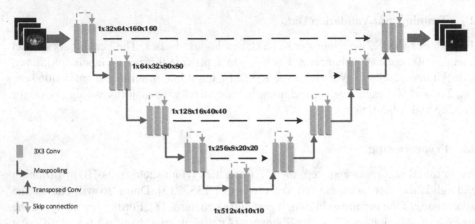

3X3 Conv

Maxpooling

Transposed Conv

Skip connection

Fig. 2. Our proposed ResU-Net3D network architecture

Next, we embed our proposed network architecture in the kidney segmentation stage. The kidney segmentation network predicts the renal contour. Prior to the tumor segmentation, the predicted renal is cropped and enhanced. The cropping step is essential to filter out all unnecessary information, such as the outer side of the kidney contours. The cropping is achieved by multiplying the input image with the prediction mask of predicted renal (output of stage 1).

Stage 2: Tumor and cyst segmentation

The tumor and cyst segmentation network take the two-channels input image by concatenating the predicted/segmented renal from stage 1 and enhanced renal after stage 1. On the z-axis, we crop the 3D cube with length 64 and stride 16 by using the center crop on the x and y-axis to clip the next cube, which comprises a single kidney with shape size (2,64,160,160) and provide that as input to the tumor and cyst segmentation network. Note that we utilize the same region props function to choose our clipping range. Finally, the designed network predicts tumor and cyst.

3   Results

We validated our approach using 30 KiTS Challenge CT images. Figure 3 provides the visual predictions of our proposed two-stage segmentation model.The quantitative results are reported in Table 1. Our proposed method achieved kidney dice of 0.96. Moreover, we evaluated our model with and without HE enhancement. Applying the HE processing step results in better tumor dice and cyst dice, as shown in Table 1. Table 2 shows the final KiTS21 result.

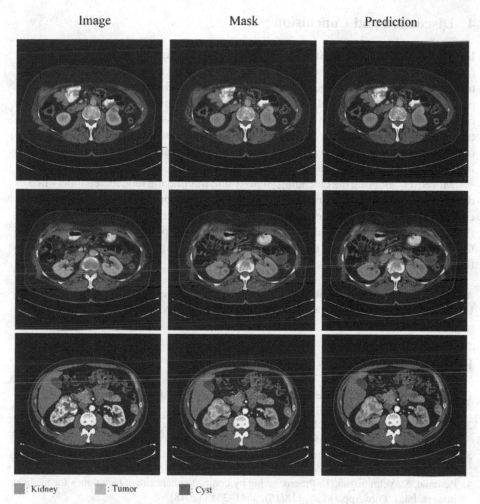

: Kidney   : Tumor   : Cyst

Fig. 3. Visual predictions of our proposed model, where the first column represents input images, column two is the ground truth (mask) images, and the third column shows the predictions of our cascaded 3D model. The colored boxes indicate each class, such as kidney, tumor, and cyst.

Table 1. The experimental outcomes on validation data with and without HE processing.

	Kidney	Tumor	Cyst
Dice with HE	0.96	0.8150	0.4504
Dice without HE		0.6710	0.4028

Table 2. The final KiTS21 results.

	Dice	Surface dice	Tumor dice
Dice with HE	0.645	0.466	0.502

128   D. Li et al.

4 Discussion and Conclusion

In this manuscript, we proposed a two-stage cascaded approach to segment kidney, renal tumor, and renal cyst. For this purpose, we employed a Residual 3DUnet architecture embedded into each stage of the cascaded pipeline. Our proposed model achieved promising segmentation results in terms of kidney and tumor segmentation. Since the boundary between the tumor and the kidney is unclear, which makes the kidney and tumor segmentation difficult. Therefore, we used histogram equalization to enhance the output from the initial stage, which serves as a second channel and enriches the image information for subsequent stage 2.

Moreover, adopting the cascading strategy and training the models separately makes distinguishing between tumor and cyst easier. Our model prediction for cyst is lower as compared to the two other segmentation tasks. As many cases have no or unclear cysts. As a result, the proposed model tends to false positives predictions. To address this in the future, we will focus on model optimization and designing complex architectures that efficiently detect cysts.

Acknowledgment. We would like to thank the KiTS21 organizers and the Shenzhen Technology University School-Enterprise Graduate Student Cooperation Fund.

References

1. Chow, W.-H., Dong, L.M., Devesa, S.S.: Epidemiology and risk factors for kidney cancer. Nat. Rev. Urol. **7**(5), 245–257 (2010)
2. Heller, N., et al.: The kits19 challenge data: 300 kidney tumor cases with clinical context, CT semantic segmentations, and surgical outcomes (2019). arXiv preprint: arXiv:1904.00445
3. Pizer, S.M., et al.: Adaptive histogram equalization and its variations. Comput. Vision Graph. Image Process. **39**(3), 355–368 (1987)
4. Perumal, S., Velmurugan, T.: Preprocessing by contrast enhancement techniques for medical images. Int. J. Pure Appl. Math. **118**(18), 3681–3688 (2018)
5. Isensee, F., Maier-Hein, K.H.: An attempt at beating the 3D U-Net (2019). arXiv preprint: arXiv:1908.02182
6. Alom, M.Z., et al.: Recurrent residual U-Net for medical image segmentation. J. Med. Imaging **6**(1), 014006 (2019)
7. He, K., et al.: Deep residual learning for image recognition. In: Proceedings of the IEEE Conference on Computer Vision and Pattern Recognition (2016)
8. https://radiopaedia.org/articles/windowing-ct

Leveraging Clinical Characteristics for Improved Deep Learning-Based Kidney Tumor Segmentation on CT

Christina B. Lund[⊠] and Bas H. M. van der Velden

Image Sciences Institute, UMC Utrecht, Utrecht University, Utrecht, The Netherlands
{C.B.Lund-2,B.H.M.vanderVelden-2}@umcutrecht.nl

Abstract. This paper assesses whether using clinical characteristics in addition to imaging can improve automated segmentation of kidney cancer on contrast-enhanced computed tomography (CT). A total of 300 kidney cancer patients with contrast-enhanced CT scans and clinical characteristics were included. A baseline segmentation of the kidney cancer was performed using a 3D U-Net. Input to the U-Net were the contrast-enhanced CT images, output were segmentations of kidney, kidney tumors, and kidney cysts. A cognizant sampling strategy was used to leverage clinical characteristics for improved segmentation. To this end, a Least Absolute Shrinkage and Selection Operator (LASSO) was used. Segmentations were evaluated using Dice and Surface Dice. Improvement in segmentation was assessed using Wilcoxon signed rank test. The baseline 3D U-Net showed a segmentation performance of 0.90 for kidney and kidney masses, i.e., kidney, tumor, and cyst, 0.29 for kidney masses, and 0.28 for kidney tumor, while the 3D U-Net trained with cognizant sampling enhanced the segmentation performance and reached Dice scores of 0.90, 0.39, and 0.38 respectively. To conclude, the cognizant sampling strategy leveraging the clinical characteristics significantly improved kidney cancer segmentation. The model was submitted to the 2021 Kidney and Kidney Tumor Segmentation challenge.

Keywords: Kidney cancer · Deep learning · Cognizant sampling · Clinical characteristics · Automated semantic segmentation

1 Introduction

According to World Health Organization a total of 431,288 people were diagnosed with kidney cancer in 2020. This makes kidney cancer the 14th most common cancer worldwide [1]. Although the number of new cases is relatively high, many patients present asymptomatic until the cancer has metastasized, and more than fifty percent of all cases are thus discovered incidentally on abdominal imaging examinations performed for other purposes [2]. Masses suspected of malignancy are investigated predominantly with contrast-enhanced computed tomography (CT) or magnetic resonance imaging (MRI) [2]. Information about size, location,

© Springer Nature Switzerland AG 2022
N. Heller et al. (Eds.): KiTS 2021, LNCS 13168, pp. 129–136, 2022.
https://doi.org/10.1007/978-3-030-98385-7_17

and morphology of the tumor can enhance treatment decisions, but manual evaluation of the CT scans remains laborious work. Scoring systems such as the R.E.N.A.L Nephrometry Score and PADUA exist to steer manual evaluation [3,4], but are subject to interobserver variability [5].

Treatment of localized kidney cancer consists of surgery of tumor and immediate surroundings (i.e., partial nephrectomy), surgery of tumor and entire kidney (i.e., radical nephrectomy), or active surveillance in case of patients who do not undergo surgery immediately but are carefully followed and evaluated for signs of disease progression [2].

Computer decision-support systems have potential to personalize treatment. Examples of such systems include volumetric measurements and radiomics approaches [6]. A crucial first step in these systems is to accurately identify kidney and kidney cancer.

The 2021 Kidney and Kidney Tumor Segmentation Challenge (KiTS21) provides a platform for researchers to test software dedicated to segmenting kidney cancer. KiTS21 does not only include images and corresponding annotations, but also an extensive set of clinical characteristics. The organizers of KiTS19 investigated whether imaging and clinical characteristics affected segmentation performance and found that tumor size had a significant association with tumor Dice score [7]. Therefore, it is reasonable to assume there may be other clinical characteristics that can be leveraged to improve segmentation.

The aim of our study was to assess whether using clinical characteristics in addition to imaging can improve the segmentation of kidney cancer.

2   Materials and Methods

We compared two different strategies (Fig. 1). As baseline, we used a 3D U-Net. We propose to improve this baseline by investigating in the validation set which clinical characteristics affect the model's performance and leverage this for cognizant sampling.

2.1   Training and Validation Data

Our submission exclusively used data from the official KiTS21 training set. The dataset contained contrast-enhanced preoperative CT scans of 300 patients who underwent partial or radical nephrectomy between 2010 and 2020. Each CT scan was independently annotated by three annotators for each of the three semantic classes: Kidney, Tumor, and Cyst. To create plausible complete annotations for use during evaluation, the challenge organizers generate groups of sampled annotations. Across these groups, none of the samples have overlapping instance annotations. It is therefore possible to compare and average them without underestimating the interobserver disagreement.

We randomly divided the dataset into training, validation, and test sets, consisting of 210, 60, and 30 patients respectively.

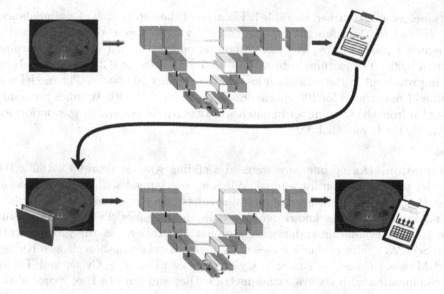

Fig. 1. By leveraging the performance of our baseline 3D U-Net model on the validation set (top row), we propose a cognizant sampling strategy based on clinical characteristics for improved segmentation

2.2 Preprocessing

Images in the dataset were acquired from more than 50 referring medical centers, leading to various acquisition protocols and thus notable differences in the image resolutions. The in-plane resolution ranged from 0.44 mm to 1.04 mm while the slice thickness ranged from 0.5 mm to 5.0 mm. To alleviate these differences, we chose to resample all images to a common resolution of 3 mm × 1.56 mm × 1.56 mm, which is the median slice thickness and twice the median in-plane resolution. Images were resampled using Lanczos interpolation, annotations were resampled using nearest neighbor interpolation. Resampling yielded a median image size of 138 × 256 × 256 voxels.

We truncated the image intensities to the 0.5 to 99.5 percentiles of the intensities of the annotated voxels in the training set. Afterwards, we performed zero-mean-unit-variance standardization based on these voxels.

Augmentations included adjustments of gamma, contrast, and brightness, addition of Gaussian noise, Gaussian blurring, scaling, rotation, and mirroring. To accommodate GPU memory limitations we cropped the images into patches of 96 × 160 × 160 voxels.

2.3 Baseline 3D U-Net

Training. The baseline 3D U-Net consisted of a downsampling path followed by an upsampling path. Downsampling was performed by max pooling operations while upsampling was done with transposed convolutions. The different parameters are described in Table 1. Because of the substantial class imbalance, we defined the loss function as the equally weighted sum of the Dice and a weighted Cross Entropy. We used Adam as optimizer with an initial learning rate of 0.005. The learning rate was reduced with a factor 0.3 if there had been no improvement in the validation loss during the last 10 epochs. The model was trained from scratch for 100 epochs. Each epoch included 400 volumes randomly sampled from the training set in batches of two. All deep learning was performed using PyTorch version 1.5.0.

Evaluation. During inference we used a sliding window (size $96 \times 160 \times 160$ voxels) to cover the entire volume. Windows overlapped with half the window size. Postprocessing consisted of retaining the two largest connected components using anatomical prior knowledge. We evaluated the model's predictions using the KiTS21 evaluation script with sampled annotations as the ground truth (see Sect. 2.1). This evaluation uses three hierarchical evaluation classes: Kidney and Masses (Kidney + Tumor + Cyst), Masses (Tumor + Cyst), and Tumor in combination with six evaluation metrics: Dice and Surface Dice scores of the three classes [8].

2.4   Cognizant Sampling Leveraging Clinical Characteristics

To devise a cognizant sampling strategy, we investigated the effect of clinical characteristics on the model's performance on the validation set. We investigated all clinical characteristics that had complete cases (i.e., no missing data) and had contrast between the patients (i.e., the variable was not the same value for all patients).

The Least Absolute Shrinkage and Selection Operator (LASSO) was used to assess which characteristics were significantly associated with kidney tumor Dice. The LASSO uses L1 regularization, which has the advantage that a sparse subset of characteristics was selected [9]. Clinical characteristics were normalized before LASSO analysis, LASSO used 5-fold cross validation.

The characteristics associated with kidney tumor Dice were weighted by the inverse of the frequency of those characteristics in the cognizant sampling strategy. For example, if smoking history was associated with kidney tumor Dice and 50% of the patients in the training population smoked, the weights of the non-smoker subset was set twice as large during cognizant sampling.

The model was retrained with no other changes than the application of the cognizant sampling strategy.

Table 1. Network description

Layer name	Layer description	Output dimension
Input	Input	$1 \times 96 \times 160 \times 160$
Dconv1	Double convolution block: 2× (3D convolution - instance normalization - ReLU activation) convolution kernel size: $3 \times 3 \times 3$, stride: $1 \times 1 \times 1$, padding: 1	$24 \times 96 \times 160 \times 160$
Mpool	Downsampling, level 1 Max pooling kernel size: $2 \times 2 \times 2$, stride: $2 \times 2 \times 2$	$24 \times 48 \times 80 \times 80$
Dconv2	Double convolution block	$48 \times 48 \times 80 \times 80$
Mpool	Downsampling, level 2	$48 \times 24 \times 40 \times 40$
Dconv3	Double convolution block	$96 \times 24 \times 40 \times 40$
Mpool	Downsampling, level 3	$96 \times 12 \times 20 \times 20$
Dconv4	Double convolution block	$192 \times 12 \times 20 \times 20$
Mpool	Downsampling, level 4	$192 \times 6 \times 10x10$
Dconv5	Double convolution block	$384 \times 6 \times 10 \times 10$
Tconv4	Upsampling, level 4 transposed convolution kernel size: $2 \times 2 \times 2$, stride: $2 \times 2 \times 2$	$192 \times 12 \times 20 \times 20$
Concat	Concatenation: [Dconv4, Tconv4]	$384 \times 12 \times 20 \times 20$
Dconv6	Double convolution block	$192 \times 12 \times 20 \times 20$
Tconv3	Upsampling, level 3	$96 \times 24 \times 40 \times 40$
Concat	Concatenation: [Dconv3, Tconv3]	$192 \times 24 \times 40 \times 40$
Dconv7	Double convolution block	$96 \times 24 \times 40 \times 40$
Tconv2	Upsampling, level 2	$48 \times 48 \times 80 \times 80$
Concat	Concatenation: [Dconv2, Tconv2]	$96 \times 48 \times 80 \times 80$
Dconv8	Double convolution block	$48 \times 48 \times 40 \times 40$
Tconv1	Upsampling, level 1	$24x96 \times 160 \times 160$
Concat	Concatenation: [Dconv1, Tconv1]	$48 \times 96 \times 160 \times 160$
Dconv9	Double convolution block	$24 \times 96 \times 160 \times x160$
Output	3D Convolution and softmax activation convolution kernel size: $1 \times 1 \times 1$, stride: $1 \times 1 \times 1$, padding: 0	$4 \times 96 \times 160 \times 160$

2.5 Statistical Evaluation

We evaluated segmentation performance (i.e., (Surface) Dice scores) of the baseline model and the model with the cognizant sampling on the test set (N = 30 patients). Normality of these performance scores was assessed using the Shapiro-Wilk test. Statistical differences in performance were assessed using the paired t-test in case of normal distributions and using the Wilcoxon signed ranked test in case of non-normal distributions. A P-value below 0.05 was considered statistically significant. All statistical analyses were performed using R version 3.6.1.

3   Results

Output of the LASSO showed that presence of chronic kidney disease, a history of smoking, larger tumor size, and radical nephrectomy instead of partial nephrectomy yielded higher tumor Dice scores (Table 2, Fig. 2).

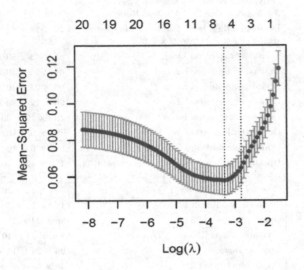

Fig. 2. Least Absolute Shrinkage and Selection Operator (LASSO) analysis shows that four variables are associated with kidney tumor Dice at one standard error from the minimum.

Table 2. The four variables associated with kidney tumor Dice in the validation set according to the Least Absolute Shrinkage and Selection Operator (LASSO)

Variable	Coefficient
Intercept	0.108
Comorbidities: chronic kidney disease	0.118
Smoking history: previous smoker	0.076
Radiographic size	0.065
Surgical procedure: radical nephrectomy	0.050

The cognizant sampling strategy significantly improved the model's segmentation performance (Table 3, Fig. 3).

The model was submitted to the 2021 Kidney and Kidney Tumor Segmentation challenge in which it was ranked 25th. The model achieved an average Dice score of 0.492, an average Surface Dice score of 0.367, and an average tumor Dice score of 0.253 on the test set.

Table 3. Dice and Surface Dice scores for the model trained with random sampling and the model trained with the proposed cognizant sampling strategy. SD = standard deviation, PR = percentile range.

Dice	Kidney	Masses	Tumor
Random sampling			
Mean (SD)	0.90 (0.08)	0.29 (0.28)	0.28 (0.29)
Median (25–75 PR)	0.93 (0.88–0.94)	0.17 (0.03–0.55)	0.15 (0.03–0.57)
Cognizant sampling			
Mean (SD)	0.90 (0.12)	0.39 (0.32)	0.38 (0.34)
Median (25–75 PR)	0.95 (0.92–0.95)	0.42 (0.07–0.70)	0.37 (0.03–0.71)
P-value	0.004	0.013	0.033
Surface dice	Kidney	Masses	Tumor
Random sampling			
Mean (SD)	0.74 (0.14)	0.12 (0.11)	0.11 (0.11)
Median (25–75 PR)	0.79 (0.66–0.84)	0.07 (0.03–0.18)	0.06 (0.03–0.18)
Cognizant sampling			
Mean (SD)	0.78 (0.16)	0.23 (0.18)	0.23 (0.21)
Median (25–75 PR)	0.84 (0.73–0.90)	0.19 (0.05–0.38)	0.22 (0.01–0.39)
P-value	0.003	<0.001	<0.001

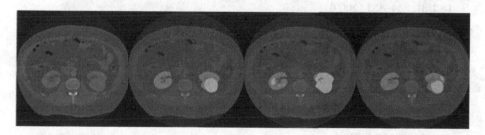

Fig. 3. Example of a 66 year old patient from the test set in whom the cognizant sampling leveraging clinical characteristics improved segmentation of the cancer. From left to right: Image, ground truth annotations, segmentation from baseline 3D U-Net, segmentation from the model using cognizant sampling

4   Discussion and Conclusion

A cognizant sampling strategy leveraging clinical characteristics significantly improved segmentation of kidney cancer on contrast-enhanced CT.

A baseline 3D U-Net was trained using random sampling. The clinical characteristics that were most associated with segmentation performance were identified using LASSO regression and used in a cognizant sampling strategy thereby leveraging the effect of the identified clinical characteristics. Previous studies showed that data-driven weighting can yield results that are independent of clinical characteristics [10,11]. Such approaches have the potential to eliminate bias towards characteristics such as smoking history, but potentially also undesirable bias such as in gender or race.

Our baseline model was a standard 3D U-Net instead of e.g. the nnU-Net provided by the challenge organizers. Since the aim of our study was to assess whether using clinical characteristics in addition to imaging can improve the segmentation of kidney cancer, we chose to direct our attention towards leveraging the potential effect of the clinical data rather than focusing solely on outperforming the results obtained with nnU-Net. It is plausible that leveraging the clinical characteristics could also improve other network architectures such as nnU-Net, because it can circumvent potential biases in patient populations. To conclude, we showed that cognizant sampling leveraging clinical characteristics improves segmentation of kidney cancer on contrast-enhanced CT.

References

1. International Agency for Research on Cancer (World Health Organization), "Kidney: Globocan 2020 - The Global Cancer Observatory," Globocan 2020, vol. 419, pp. 1–2 (2020)
2. Ljungberg, B., et al.: EAU guidelines on renal cell carcinoma: 2014 update. Eur. Urol. **67**(5), 913–924 (2015)
3. Kutikov, A., Uzzo, R.G.: The RENAL nephrometry score: a comprehensive standardized system for quantitating renal tumor size, location and depth. J. Urol. **182**(3), 844–853 (2009)
4. Ficarra, V., et al.: Preoperative Aspects and Dimensions Used for an Anatomical (PADUA) classification of renal tumours in patients who are candidates for nephron-sparing surgery. Eur. Urol. **56**(5), 786–793 (2009)
5. Spaliviero, M.: Interobserver variability of RENAL, PADUA, and centrality index nephrometry score systems. World J. Urol. **33**(6), 853–858 (2014). https://doi.org/10.1007/s00345-014-1376-4
6. Ursprung, S., et al.: Radiomics of computed tomography and magnetic resonance imaging in renal cell carcinoma-a systematic review and meta-analysis. Eur. Radiol. **30**, 3558–3566 (2020)
7. Heller, N., et al.: The state of the art in kidney and kidney tumor segmentation in contrast-enhanced CT imaging: Results of the KiTS19 challenge. Med. Image Anal. **67**, 10182 (2021)
8. Nikolov, S., et al.: Deep learning to achieve clinically applicable segmentation of head and neck anatomy for radiotherapy, CoRR (2018)
9. Tibshirani, R.: Regression shrinkage and selection via the lasso. J. Roy. Stat. Soc. Ser. B (Methodol.) **58**(1), 267–288 (1996)
10. Van Der Velden, B.H., et al.: Complementary value of contralateral parenchymal enhancement on DCE-MRI to prognostic models and molecular assays in high-risk ER-positive/HER2-negative breast cancer. Clin. Cancer Res. **23**(21), 6505–6515 (2017)
11. van der Velden, B.H.M., Sutton, E.J., Carbonaro, L.A., Pijnappel, R.M., Morris, E.A., Gilhuijs, K.G.A.: Contralateral parenchymal enhancement on dynamic contrast-enhanced MRI reproduces as a biomarker of survival in ER-positive/HER2-negative breast cancer patients. Eur. Radiol. **28**(11), 4705–4716 (2018). https://doi.org/10.1007/s00330-018-5470-7

A Coarse-to-Fine 3D U-Net Network for Semantic Segmentation of Kidney CT Scans

Yasmeen George[✉]

School of Information Technology, Monash University, Subang Jaya, Malaysia
yasmeen.george@monash.edu

Abstract. The number of kidney cancer patients is increasing each year. Computed Tomography (CT) scans of the kidneys are useful to assess tumors and study tumor morphology. Semantic segmentation techniques enable the identification of kidney and surrounding anatomy on the pixel level. This allows clinicians to provide accurate treatment plans and improve efficiency. The large size of CT volumes poses challenges for deep segmentation methods as it cannot be accommodated on a single GPU in its original resolution. Downsampling CT scans influences the segmentation performance. In this paper, we present a coarse-to-fine cascaded network based on 3D U-Net architecture for semantic segmentation of kidney CT volumes into three classes kidney, tumor, and cyst. A two stage approach is implemented where a 3D U-Net model is first trained on downsampled CT volumes to delineate kidney region. This is followed by another 3D U-Net model which is trained using the full resolution images cropped around the areas of interest generated by first stage segmentation results. A set of 300 CT scans were used for training and evaluation. The proposed approach scored 0.9748, 0.8813, 0.8710 average dice for kidney, tumor and cyst using 3D cascade U-Net model. The performance of the cascade network outperformed other trained U-Net models based on 2D, 3D low resolution and 3D full resolution. The model also achieved the 3rd place in the leaderboard of KiTS21 challenge with a mean sampled average dice score of 0.8944 and a mean sampled average surface dice score of 0.8140 using a test set of 100 CT scans.

Keywords: Semantic segmentation · Cascaded network · 3D U-Net · Medical image diagnostics

1 Introduction

The number new cases with kidney tumors is increasing each year [2]. Globally, kidney cancer is the sixth most commonly diagnosed cancer for men and the ninth for women [1]. In Australia, it is the seventh most common cancer [2]. Kidney tumors can be classified as benign, indolent or malignant. Benign tumors can grow slowly but does not spread to other tissues. A malignant tumor is cancerous

© Springer Nature Switzerland AG 2022
N. Heller et al. (Eds.): KiTS 2021, LNCS 13168, pp. 137–142, 2022.
https://doi.org/10.1007/978-3-030-98385-7_18

with renal cell carcinoma is the most common type of kidney cancer leading to 140,000 deaths annually worldwide [4]. An indolent tumor is also cancerous, but this type of tumor rarely spreads to other parts of the body. A great research effort is being invested on studying the relationship between tumor morphology (size, shape and appearance) and surgical outcomes. Small tumors are often detected incidentally when the patient has a scan for an unrelated problem [3]. There is a need for automated semantic segmentation and classification methods to objectively quantify the severity of kidney tumors in order to better inform treatment decisions. This will also help doctors to solve diagnostic problems and improve efficiency.

Recent advances in imaging tools facilitate the detection and diagnosis of kidney tumors and contribute to preventive treatment of kidney cancer. Clinicians predominantly rely on imaging tests, primarily Computed Tomography (CT), to both diagnose and stage renal cell carcinomas [2]. CT uses x-rays to provide cross-sectional images of the body from different angles. The 2D slices are combined to form the final 3D volume for the kidney. CT scan reveals any abnormalities or tumors and can be used to measure the size of the tumor.

Semantic segmentation plays an important role in diagnostics support systems in the medical domain. It enables the identification of different objects in images on the pixels level. In this paper, we apply semantic segmentation for kidney tumor diagnosis where each voxel in the CT scan is labeled as background, kidney, renal tumors or cyst.

2   Methods

Inspired by nnU-Net work [3], we implemented a coarse-to-fine cascaded U-Net approach which has two stages. In the first stage, a 3D U-Net model is trained on downsampled images to roughly delineate kidney region. In the second stage, a 3D U-Net model is trained to have more detailed segmentation of the three classes (kidney, tumor, cyst) using the full resolution images guided by the first stage segmentation maps (See Fig. 1 for the network architecture).

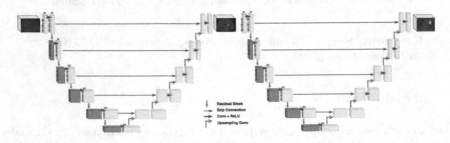

Fig. 1. Kidney cancer segmentation framework using 3D cascaded U-Net architecture

2.1 Training and Validation Data

In this paper, we use the kits21 dataset [1]. The KiTS21 challenge organizers have produced ground truth semantic segmentations for arterial phase abdominal clinical CT scans for training and validation. The dataset consists of 300 unique kidney cancer patients who underwent partial or radical nephrectomy for suspected renal malignancy between 2010 and 2020 at either an M Health Fairview or Cleveland Clinic medical center. Each CT volume consists of 29–1059 slices of 512–796 × 512 pixels. The voxel dimensions are [0.44–1.04, 0.44–1.04, 0.5–5] mm.

Regions in CT scan are for 3 different classes, kidney, tumor and cyst. The ground truth segmentation mask is provided for each class separately. A group of trainees that consists of medical students, undergraduates planning to study medicine, and one Computer Science PhD student annotate region of interests by placing 3D bounding boxes as well as guidance pins for axial slices. A group of experts that include radiologists and urologic cancer surgeons review the annotations and leave comments if necessary. The round review is repeated until experts approve the annotation for accuracy and completeness. Finally the guidance for each region is sent individually to three laypeople for contour annotations. Trainees review the contour annotations and ensure that they adhere to the expert-approved guidance. Finally, contour annotations are postprocessed to generate segmentations.

All DL models in this paper used majority aggregation based ground truth segmentation for training and validation (Fig. 2).

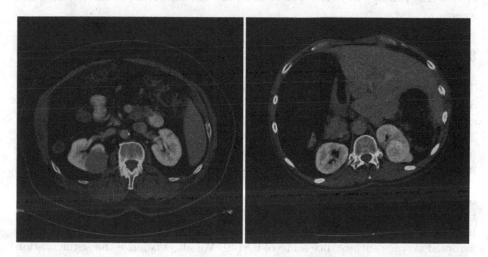

Fig. 2. Sample 2D slices from 2 CT scans with annotated kidney, tumor and cyst in blue, orange and green in order (Color figure online)

2.2   Data Preprocessing

The CT intensities (HU) were transformed by subtracting mean and dividing by standard deviation. The data augmentation methods include random rotations, gamma transformation, and random cropping.

2.3   Proposed Method

The proposed approach is based on two stage cascaded network for kidney, tumor and cyst segmentation using 3D U-Net architecture. In the first stage, each CT scan was resampled using third order spline interpolation to a spacing of $1.99 \times 1.99 \times 1.99$ mm resulting in median volume dimensions of $207 \times 201 \times 201$ voxels. While in the second stage, a spacing of $0.78 \times 0.78 \times 0.78$ mm was used with median volume dimensions of $528 \times 512 \times 512$ voxels.

The 3D U-Net architecture had an encoder and a decoder path each with five resolution steps. The encoder part was performed using strided convolutions starting with 30 feature maps then doubling up each level to a maximum of 320. The decoder part was based on transposed convolutions. Each layer consists 3D convolution with $3 \times 3 \times 3$ kernel and strides of 1 in each dimension, leaky ReLU activations, and instance normalization.

All the models were trained from scratch using 5-fold cross-validation with a patch size of $128 \times 128 \times 128$ that was randomly sampled from the input resampled volumes. The models were trained using stochastic gradient descent (SGD) optimizer for 1000 epochs using a batch of size 2 with 250 batches per epoch. The training objective was to minimize the sum of cross-entropy and dice loss.

3   Results

The proposed models are implemented using nnU-Net framework [3] with Python 3.6 and PyTorch framework on NVIDIA Tesla V100 GPUs. For performance evaluation, we report the average 5-fold dice coefficient and Surface Dice (SD). To compare the performance of the proposed network, we report the performance measures for 2d U-net model, 3D full resolution and 3D low resolution U-Net models. Table 1 displays the dice and SD scores for all trained models. The table shows that the dice scores of 0.9748, 0.8813, 0.8710 for kidney, tumor and cyst in order. Another test set which consisted of 100 CT scans was used on to evaluate the submitted solutions on KiTS21 challenge. Our model achieved the 3rd place in the leaderboard with a mean sampled average dice score of 0.8944 and a mean sampled average surface dice score of 0.8140. We also visualize the segmentation results for all trained models in Fig. 3.

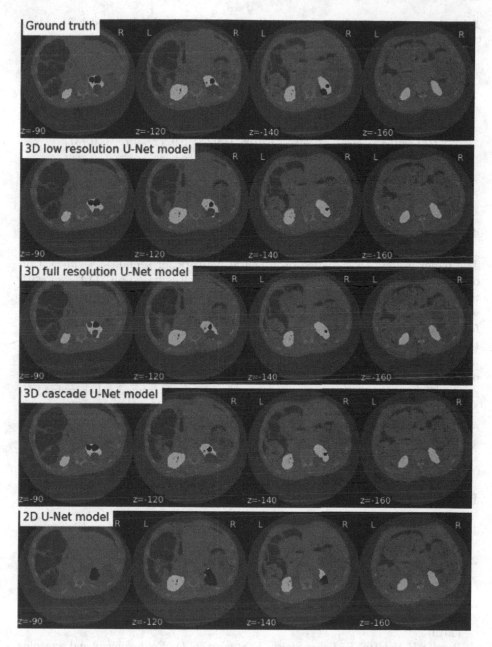

Fig. 3. Segmentation results for kidney, tumor and cyst using U-Net trained models

Table 1. Average 5-fold performance measures for the trained U-Net models for kidney tumor segmentation

U-Net model	Dice scores			SD scores		
	Kidney	Cyst	Tumor	Kidney	Cyst	Tumor
2D	0.9615	0.7686	0.7367	0.9072	0.6154	0.5787
3D FullRes	0.9691	0.8693	0.8535	0.9359	0.7626	0.7415
3D LowRes	0.9685	0.8705	0.8568	0.9272	0.7508	0.7377
3D Cascade	0.9748	0.8813	0.8710	0.9448	0.7728	0.7605
KiTs21 Leaderboard	0.8944			0.8140		

4 Discussion and Conclusion

The paper presents a cascaded deep neural network for semantic segmentation of kidneys and surrounding anatomy. The approach is based on 3D U-Net with two stages for training. The first stage model is trained on downsampled volumes while the second stage model is trained on the cropped region of interest in its full resolution. The models were able to accurately segment the kidney and tumor while the performance of cyst segmentation is lowest due to the small number of annotated cases with cyst. The model achieved an average dice score of 0.9748, 0.8813, 0.8710 for kidney, tumor and cyst using 3D cascade U-Net model. The performance of the cascade network outperformed other trained U-Net models based on 2D, 3D low resolution and 3D full resolution. The model also achieved the 3rd place in the leaderboard of KiTS21 challenge with a mean sampled average dice score of 0.8944 and a mean sampled average surface dice score of 0.8140 using a test set of 100 CT scans.

References

1. The 2021 kidney and kidney tumor segmentation challenge. https://kits21.kits-challenge.org
2. Heller, N., et al.: The state of the art in kidney and kidney tumor segmentation in contrast-enhanced CT imaging: results of the kits19 challenge. Med. Image Anal. **67**, 101821 (2021)
3. Isensee, F., Petersen, J., Kohl, S.A.A., Jäger, P.F., Maier-Hein, K.: nnU-Net: Breaking the spell on successful medical image segmentation. ArXiv abs/1904.08128 (2019)
4. Rossi, S.H., Klatte, T., Usher-Smith, J., Stewart, G.D.: Epidemiology and screening for renal cancer. World J. Urol. **36**(9), 1341–1353 (2018). https://doi.org/10.1007/s00345-018-2286-7

3D U-Net Based Semantic Segmentation of Kidneys and Renal Masses on Contrast-Enhanced CT

Mingyang Zang[1], Artur Wysoczanski[1], Elsa Angelini[1,2], and Andrew F. Laine[1(✉)]

[1] Department of Biomedical Engineering, Columbia University, New York, NY, USA
{mz2846,al418}@columbia.edu
[2] ITMAT Data Science Group, NIHR Imperial BRC, Imperial College, London, UK

Abstract. The accurate, automated detection and segmentation of renal tumors is of great interest for the imaging-based diagnosis, histologic subtyping, and management of suspected renal malignancy. The KiTS21 Grand Challenge provides 300 contrast enhanced CT images with kidney, tumors and cysts with corresponding manual annotation, to facilitate the development of robust segmentation algorithms for this task. In this work, we present an adaptation of the historically-successful 3D U-Net architecture, combined with deep supervision, foreground oversampling and large-scale image context, and trained on the majority-prediction segmentation masks. Our model achieved test-set performance of 97.0%, 85.1%, and 81.9% volumetric Dice score, and 93.7%, 72.0%, and 70.0% surface Dice score, on combined foreground, renal masses, and renal tumors, respectively, which tied for sixth place among challenge participants.

Keywords: 3D U-Net · Medical image segmentation · Renal tumor detection

1 Introduction

With the increasing quantity and quality of volumetric medical images, deep-learning methods are gaining in popularity, and equal or exceed the performance of expert human reviewers on a wide variety of detection, segmentation and classification tasks [8]. Automated tumor detection and delineation is of particular interest in the context of renal masses, which are often incidentally detected and whose imaging features have significant implications for patient management [6].

3-D encoder-decoder networks, such as 3D U-Net [9] and V-Net [5], have been shown to be robust to a wide variety of segmentation tasks across imaging modalities and protocols. Crucially for medical imaging applications, where the amount of training data is often severely limited due to privacy concerns and time- and cost-prohibitive annotation, such networks can generally be trained end-to-end from very few images. Although the U-Net architecture has subsequently been

N. Heller et al. (Eds.): KiTS 2021, LNCS 13168, pp. 143–150, 2022.
https://doi.org/10.1007/978-3-030-98385-7_19

augmented by the incorporation of residual blocks, attention gating, and other features, the "vanilla" U-Net often outperforms its successors. Motivated by this observation, nnU-Net (short for "no new U-Net") [4] focuses primarily on standard network architectures, while tuning hyperparameters such as batch size, optimizer parameters, and patch and kernel size to improve the network generalization ability. The nnU-Net has achieved top performance by mean Dice similarity on all but one class of the Medical Segmentation Decathlon challenge [1], and was also the basis for the top-performing entry in the KiTS19 Grand Challenge [2,3]. Medical images are mostly single channeled and less diverse than natural images [7]. Hence, we hypothesize that a straightforward 3D U-Net architecture with proper processing and sampling of original data and minor modifications to network architecture, may achieve similarly high performance on the current challenge dataset.

2   Methods

2.1   Training and Validation Data

Our submission made use of the official KiTS21 training set alone. The data was trained with 240 cases and validated using the remaining 60 cases. Network training and validation are performed on the majority-voting segmentation masks provided by the challenge organizers.

2.2   Preprocessing

The original CT scans are resampled to isotropic 1.99 mm × 1.99 mm × 1.99 mm resolution by third-order spline interpolation, and the ground-truth segmentation masks are resampled using nearest-neighbor interpolation.

We achieved optimal network performance by performing case-by-case clipping of Hounsfield intensity values to the 0.5th and 99.5th percentile, after which each volume was normalized by subtracting the mean intensity and dividing by the standard deviation. All results reported in this paper were obtained with this normalization technique. Data augmentation including mirroring, rotation about all axes, brightness control, gamma correction, contrast adjustment, and scaling were applied randomly at run time to all image patches, with a probability of 0.4 for each operation.

2.3   Network Architecture

The network is constructed based on the 3D U-Net [9] architecture, using the nnUNet framework; the network architecture is depicted in Fig. 1. The breadth of first convolutional block is set to 32 channels, and doubles at each downsampling step. Downsampling is continued until reaching output dimension 4×4×4. At the shallowest layers of the upsampling arm, we generate downsampled prediction masks by convolution followed by softmax output. Such supervision enables the network to predict correctly starting from the low resolution and avoid passing wrong segmentation information to the higher resolution layers.

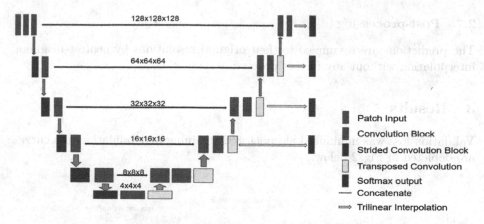

Fig. 1. The scheme of the 3D U-Net. All convolution blocks use $3 \times 3 \times 3$ kernels and the transposed convolution use $2 \times 2 \times 2$ kernels.

2.4   Loss Function

We use the sum of categorical cross-entropy and soft Dice loss as our objective function at each output layer. The final loss function is the weighted sum of the losses calculated at each output layer of the network, with the weight decreasing by a factor of two with each drop in resolution.

2.5   Optimization Strategy

We choose a patch size of $128 \times 128 \times 128$ and set the batch size to 4, maximizing the patch volume under the constraints imposed by GPU memory. Patches are sampled at random from the training images at run time, with oversampling of the foreground classes achieved by requiring that at least one-third of each patch be occupied by a foreground label (kidney, cyst, or tumor) to focus network training on the foreground classes. We implement SGD with an initial learning rate of 0.06, 0.99 momentum and Nesterov as our optimizer. We define one epoch as 250 batches and the whole training phase lasts for 1000 epochs.

2.6   Validation

Our hold-out validation set consists of 60 cases selected at random without replacement from the public dataset. Validation is performed using a sliding window approach with a stride equal to half the patch size.

2.7 Post-processing

The predictions are resampled to their original resolutions by nearest-neighbor interpolation without any further processing.

3 Results

Validation loss was minimized at epoch 892; training and validation loss curves are depicted in Fig. 2 below.

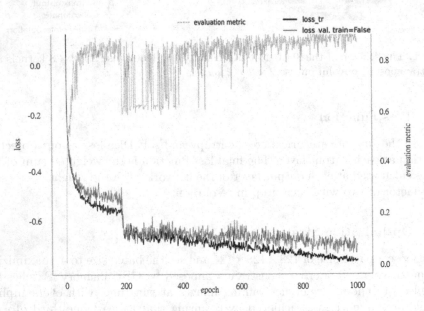

Fig. 2. Training loss (blue), validation loss (red), and exponential moving average of the composite Dice (green) across 1000 epochs. (Color figure online)

Model selection was performed by comparison to the majority-vote segmentation on the validation dataset; performance metrics for the final network were calculated against both the majority-vote and individual annotations for comparison. On the test set, the network attained a mean volumetric Dice score of 0.970, 0.851 and 0.819 on the "kidney + masses", "masses" and "tumor" classes, respectively; mean surface Dice scores on the same classes reached 0.937, 0.720 and 0.700 respectively. Performance metrics for both the validation and test sets are shown in Table 1 below, with representative slices from the best-performing validation cases presented in Fig. 3.

Table 1. Volume Dice and surface Dice on the validation and test sets.

	Dice (Majority/Individual/Test)	Surface Dice
Kidney + Masses	0.971/0.963/0.970	0.937/0.919/0.937
Masses	0.861/0.856/0.851	0.762/0.749/0.720
Tumor	0.840/0.835/0.819	0.743/0.729/0.700

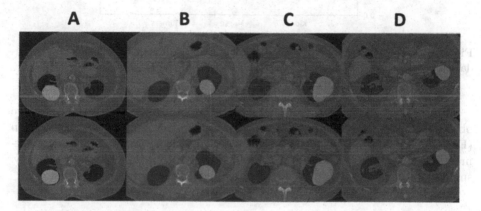

Fig. 3. Representative segmentation from the top-four validation cases, as determined by the mean of volumetric and surface Dice scores across all evaluation classes. Kidneys are annotated in red, renal tumors in green, and cysts in blue. Top row: predicted segmentation; Bottom row: majority-vote ground truth. (Color figure online)

Both the mean Dice coefficient and mean surface Dice are relatively robust to the choice of majority or individual annotation for evaluation, with a maximum absolute decrease of 0.008 for Dice coefficient and 0.018 for surface Dice across all hierarchical classes when comparing individual annotation to the majority-voting scheme. Validation performance under the two evaluation methods was also strongly correlated at the subject level (Spearman r = 0.996). The network generalized well to the test set, with an absolute decrease of 0.016 and 0.029 in volume Dice and surface Dice coefficients, respectively, on the tumor class.

The distributions of volumetric and surface Dice coefficients across the validation set (Fig. 4) are left-skewed with multiple outliers at low Dice coefficient, most prominently among masses and tumors. Among validation cases containing a solitary renal tumor, we observed that all outliers with respect to volumetric and surface Dice have tumor volumes below the median (24.75 cm^3), as computed from the majority-vote segmentation (Fig. 5). Performance on tumors below the median volume differ from performance on larger tumors, both by volumetric Dice (median 0.832, vs. 0.942, Mann-Whitney U = 28, p < 10^{-9}) and surface Dice coefficients (median 0.751 vs. 0.829, Mann-Whitney U = 216, p < 0.005).

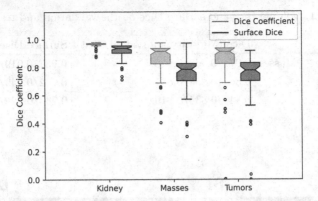

Fig. 4. Box-and-whisker plots for the volumetric and surface Dice coefficients of validation cases, for all evaluation classes.

Among cases in the validation set with worst tumor Dice coefficient, we observe three general failure modes for both tumor and cyst segmentation (Fig. 6): omission or false detection of small masses (Fig. 6A, D), under- segmentation (Fig. 6B, C) and tumor/cyst mis-classification (Fig. 6D, E). Generally, small cysts were particularly vulnerable to under-segmentation or omission.

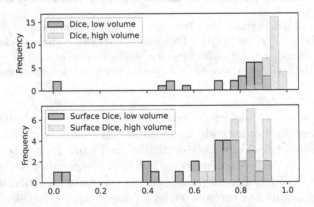

Fig. 5. Histograms of tumor volumetric (top) and surface (bottom) Dice coefficients, in validation cases with a single renal tumor (n = 57). For illustrative purposes, we define the "low volume" class to include tumors with majority-vote volume below the median on this dataset, with "high volume" containing the remainder.

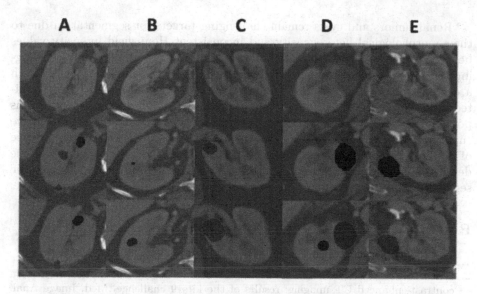

Fig. 6. Representative examples of segmentation failure, defined by lowest volumetric Dice score for the tumor class. Kidneys are annotated in red, renal tumors in green, and cysts in blue. Top row: CT image; Middle row: predicted mask overlaid; Bottom row: majority-vote ground truth segmentation overlaid. Note missed/false-positive small masses (cases A, D), under-segmented cysts and tumors (cases B, C), and mis-classified masses (cases D, E). (Color figure online)

4   Discussion and Conclusion

Our work demonstrates once again that 3D U-Net architectures achieve competitive performance on kidney and renal mass segmentation in the KiTS21 dataset. Building on the nnU-Net framework, we incorporate deep supervision, a targeted foreground-oversampling strategy, and large-volume image patches with maximized batch size to optimize network performance. On the test dataset, we achieve competitive volume Dice scores of 0.970, 0.851 and 0.819 for kidney (including tumor and cysts), mass and tumor and surface Dice scores of 0.937, 0.720 and 0.700 for kidney (including tumor and cysts), mass and tumor respectively, tying for sixth place out of 25 participating teams.

Validation performance metrics indicate that the majority-vote segmentation is a reasonable proxy for individual reviewers' annotations. Both volumetric and surface Dice scores against individual annotations are quite similar to those attained on the majority predictions used for training. Nevertheless, alternative strategies leveraging the individual annotations directly, including ensemble prediction using networks trained by separate reviewers, may be of further interest. For applications where minimizing boundary error is particularly desirable, it may be beneficial to add a proxy for the surface Dice coefficient to the objective function directly.

Renal tumors and cysts remain challenging targets for segmentation due to their morphological heterogeneity and inconsistent Hounsfield intensity values between CT scans [7]. We have found that small masses are especially challenging for our current architecture, and that low tumor volume is associated with a decrease in both volumetric and surface Dice scores. Although U-Net architectures are known to perform robustly even with limited training data, it is possible that given a larger training set, higher-capacity models may achieve superior performance in renal mass segmentation. Given the relative abundance of publicly-available contrast-enhanced CT without voxel-level annotation, the design of semi-supervised or weakly-supervised architectures for 3D semantic segmentation is of particular interest to improve upon our current performance.

References

1. Antonelli, M., et al.: The medical segmentation decathlon (2021)
2. Heller, N., et al.: The state of the art in kidney and kidney tumor segmentation in contrast-enhanced CT imaging: results of the kits19 challenge. Med. Image Anal. **67**, 101821 (2021)
3. Isensee, F., Maier-Hein, K.H.: An attempt at beating the 3D U-Net (2019)
4. Isensee, F., et al.: nnU-Net: Self-adapting framework for u-net-based medical image segmentation (2018)
5. Milletari, F., Navab, N., Ahmadi, S.A.: V-net: fully convolutional neural networks for volumetric medical image segmentation. In: 2016 Fourth International Conference on 3D Vision (3DV), pp. 565–571 (2016). https://doi.org/10.1109/3DV.2016.79
6. Wang, Z.J., Westphalen, A.C., Zagoria, R.J.: CT and MRI of small renal masses. Br. J. Radiol. **91**(1087), 20180131 (2018)
7. Zhao, W., Zeng, Z.: Multi scale supervised 3D U-Net for kidney and tumor segmentation (2019)
8. Zhou, S.K., et al.: A review of deep learning in medical imaging: imaging traits, technology trends, case studies with progress highlights, and future promises. Proc. IEEE **109**(5), 820–838 (2021). https://doi.org/10.1109/JPROC.2021.3054390
9. Çiçek, Ö., Abdulkadir, A., Lienkamp, S.S., Brox, T., Ronneberger, O.: 3D U-Net: learning dense volumetric segmentation from sparse annotation. In: Ourselin, S., Joskowicz, L., Sabuncu, M.R., Unal, G., Wells, W. (eds.) MICCAI 2016, Part II. LNCS, vol. 9901, pp. 424–432. Springer, Cham (2016). https://doi.org/10.1007/978-3-319-46723-8_49

Kidney and Kidney Tumor Segmentation Using Spatial and Channel Attention Enhanced U-Net

Sajan Gohil[✉] and Abhi Lad

Pandit Deendayal Energy University, Gandhinagar, India
{sajan.gict16,abhi.lce16}@sot.pdpu.ac.in

Abstract. Kidney and Kidney tumor segmentation from CT scans has tremendous potential to help doctors in early diagnosis and localization of tumor, its size and type and for making timely treatment plans. However, considering the nature and volume of data, it is difficult and time consuming to train on such CT scans. In this paper, we propose enhancements to the 3D U-Net model to incorporate Spatial and Channel Attention in order to improve the identification and localization of segmentation structures by learning on spatial context. When compared with Residual U-Net model with greater depth and more feature maps, our Spatial and Channel Attention enhanced U-Net with less depth and feature maps performed significantly better on validation and training set when trained under similar conditions.

Keywords: Kidney tumor · Segmentation · 3D U-Net · Spatial attention · Channel attention

1 Introduction

Manually identifying tumors from CT scans is a tedious and time-consuming process. It is also a difficult task as there can be inconsistencies in proper segmentation even by experienced practitioners. In some cases, the boundaries of lesions can also be unclear in CT scan images and the images can also have poor contrast and structure definition. To help solve these issues, many computer vision based deep learning methods have been proposed and developed which are trained to segment and/or classify such lesions. One of the most popular models for such tasks is U-Net [1] which can be modified for 3D convolutions and with residual units as proposed in [2]. Furthermore, the size of such imaging modalities can be huge, thus making it difficult to train on all the data. So proper data preprocessing and manipulation can also play an important role in making the training efficient and viable and make the model more robust to drift in data.

In this paper, we enhance the basic U-Net by including visual attention introduced in [3] with 3D convolutions. Attention blocks help our model train more effectively by refining the features with the help of a global attention map. Combination of attention mechanism with U-Net has previously been explored in [4]. [5] introduced Squeeze and

---

S. Gohil and A. Lad—Equal Contribution.

© Springer Nature Switzerland AG 2022
N. Heller et al. (Eds.): KiTS 2021, LNCS 13168, pp. 151–157, 2022.
https://doi.org/10.1007/978-3-030-98385-7_20

excite mechanism which is advantageous for channel refinement as used in works such as [6] with variations of attention mechanism across spatial and channel dimensions and [7] with project and excite blocks for segmentation of volumetric medical scans. Models with attention have previously been used for Kits19 [8] dataset in works such as [9] which use attention modules at the end of their model for final refinement of features. In this paper, we use architecture is similar to [10] which uses spatial and channel attention separately for fine and sparse features. We also trained a standard U-Net with residual blocks for comparison and found that our spatial and channel attention enhanced U-Net performed better on training and validation sets, that too with less number of epochs. We also resample our data, especially along the z-axis to effectively increase the number of training image slices per volume. In order to reduce training time on constrained resources, we use Nvidia Clara SmartCache [11] to improve training times without loading the entire dataset to cpu memory. The following sections describe in detail the data methods and the custom deep learning U-Net model used in our challenge submission.

2 Methods

In this section we discuss the data split for training, preprocessing steps and data augmentations from the point of view of generalizability, and finally the custom U-Net model along with the details about attention blocks.

2.1 Training and Validation Data

Our submission made use of the official KiTS21 training set alone. For segmentation masks, we have used "aggregated_AND" based final masks to train our model. We have refrained from using data from other similar studies and instead use data augmentation techniques to adapt the segmentation model for better generalizability. The data was split in a 90:10 ratio for training and validation respectively. The data is randomly shuffled before creating the validation split. The final submission uses a model trained on the entire training set.

2.2 Preprocessing

The number of slices for each volume sample is different and the number of slices can also be high enough to consume all available resources during training. To avoid such a scenario, we resample the data to $2 \times 1.62 \times 1.62$ mm to have the same voxel spacing across the patient image volumes. The lower spacing along the z-axis increases the number of training slices per patient and thus helps generalizability. We further clean the data by keeping only the body structures in frame by cropping the foreground. The final cropped data is further divided in chunks of $64 \times 128 \times 128$ volumes for training and evaluation. For training Residual U-Net, we divide the cropped data in chunks of $64 \times 160 \times 160$ to capture higher spatial information.

The intensity values vary based on physical properties of structures and thus we remove unnecessary values corresponding to structures like bone and air by clipping to range $(-80, 305)$. Then the clipped values are rescaled between 0 and 1 for all images.

In order to improve training times and prevent crashes by cpu memory bottleneck, we use the Nvidia Clara SmartCache dataset loader. We set the cache rate = 0.4 and replace rate = 0.5. This ensures sufficient data is cached in memory for training while replacing 50% of cached data at each step without caching the entire dataset and choking the cpu memory.

2.3 Data Augmentations

In order to make the model generalizable, we introduce variance in data using data augmentation techniques. To introduce spatial variance, we use 3D Elastic deformation with a probability of 0.5. The parameters for 3D elastic deformation are: sigma (smoothness factor) = (5, 8); magnitude = (50, 150); translate = (10, 10, 5) in pixels; rotate = (5, 5, 180) in degrees, scale = (0.1, 0.1, 0.1) in proportion of image size. This augmentation introduces variation in shape of structures while maintaining spatial information.

For introducing perceptual or intensity variations, we use random intensity shift with maximum intensity offset value of 0.1. We also use Gaussian noise with mean = 0 and standard deviation = 0.1. Both of the intensity-based augmentations are applied with probability of 0.25 individually.

2.4 Proposed Method

Following the success of U-Net models and its variants, we have decided to use the U-Net model with 3D Convolution blocks as our base architecture. Our base architecture has 3 encoding and 3 decoding blocks with a bottleneck block in between. Figure 1(A) describes the architecture used in our submission. The number of feature maps for encoder blocks are (32, 64, 128), followed by 128 feature maps for the bottleneck block. The feature maps from skip connections are stacked in decoding blocks resulting in feature maps (256, 128, 64). All the encoding and decoding blocks have kernel size of $3 \times 3 \times 3$ and stride value as 2, except for bottleneck block with kernel size of $3 \times 3 \times 3$ and stride value 1, and the final output conv layer with kernel size $1 \times 1 \times 1$ and stride value 1. To enhance this base model, we have added Spatial attention and Channel attention modules, as introduced in [12] in an architecture similar to [10] which was proposed for 2D segmentation tasks.

Spatial Attention Block: The spatial attention block is responsible for identifying where the useful information is present in the image, by utilizing the inter-spatial relationships of the image features. Figure 1(B) describes the spatial attention block used in our proposed approach. The spatial attention block uses mean and max operations along channel dimension followed by 3D conv ($7 \times 7 \times 7$) to identify region of interest and multiply the attention map with output of preceding 3D Conv + ADN (Attention + Dropout + Normalization) block to filter out location of important features.

Channel Attention Block: Instead of focusing on where the important feature is, the channel attention block identifies what is useful in a given image. Figure 1(C) describes the channel attention block used in our approach. The channel attention block uses mean and max values across spatial dimensions followed by a conv block to identify what is important in a given volume.

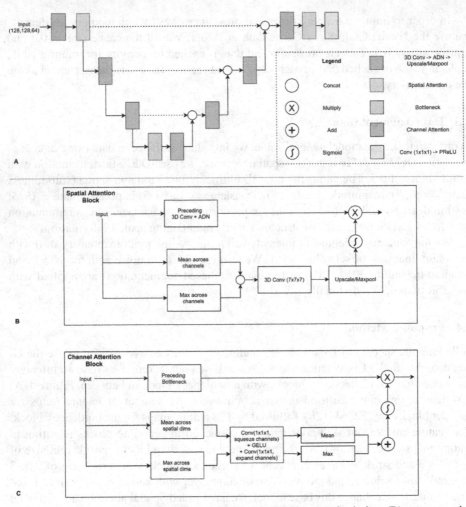

Fig. 1. (A) describes the enhanced U-Net architecture used in our submission. (B) represents the working of Spatial Attention Block. (C) represents the working of Channel Attention Block. (B) and (C) includes the interaction of attention mechanism with preceding blocks in the U-Net.

Our implementation is similar to model architecture of [10]. We have added Spatial attention to the first encoding and last decoding blocks having the largest dimensions. We have added channel attention to the bottleneck layer. The depth and number of feature maps of our enhanced U-Net has been limited due to resource constraints but can be increased to further improve the results.

2.5 Residual U-Net for Comparison

We also train a residual 3D U-Net to compare performance of our proposed model. The residual U-Net is similar to [13], however the number of features maps are different (16, 32, 64, 128, 256). The absence of attention blocks allows for a greater number of encoding-decoding blocks with more number of feature maps. Also, as compared to 64 × 128 × 128 patch size used for training our proposed model, we increase the size of patches to 64 × 160 × 160 for residual U-Net to increase the spatial information per patch.

2.6 Implementation and Training

The model is implemented in pytorch using MONAI [11] framework. Both the models output 4 channels corresponding to 4 classes including the background. For training we have used DiceCE loss which is a combination of Dice and Cross Entropy loss functions. We have used AdamW optimizer with a learning rate of $10e - 4$. Both models are trained for 300 epochs for comparison and the final submission is made using our proposed model which is trained for 500 epochs on Nvidia P100 GPU with 16 GB VRAM and 24 GB CPU RAM. Validation is performed at the end of each epoch using the overall dice score as metric. Each epoch takes ~1.5 min and the whole model completes 500 epochs in ~13 h.

2.7 Inference Procedure

For final inference, since the model is trained on 64 × 128 × 128 sized chunks of input volume, we use sliding window inference with overlap = 0.8. The high overlap value increases the inference time but also improves the segmentation results. The 4 channeled output is converted to 1 channel by using maximum probability of segmented classes. And finally, the voxel spacing of the segmented volumes are restored to original spacing of the input volume by inverse transform.

3 Results

Here we provide the comparative results of our proposed model with residual U-Net based on training and validation set. We provide overall dice and structure specific dice scores. Finally, we also provide the dice scores on the test set as provided by kits21 evaluation system. Table 1 shows the performance of our proposed model and Residual U-Net on validation and complete training set after 300 epochs. Our approach outperformed Residual U-Net by margin or ~0.2 in mean Dice score on both validation and training sets.

Figure 2 shows the segmentation results corresponding to a volume from validation set of our proposed model after 300 epochs. As it can be seen, the results do not suffer from jagged/rough edges from resampling the input volume.

The results on the test set of KiTS21 are summarized in Table 2. Our approach ranked 18th on the KiTS21 challenge leaderboard.

Table 1. Dice score for individual classes and mean across classes of our approach compared with residual U-Net on validation and complete training set after 300 epochs.

	Set	Kidney	Tumor	Cyst	Mean
Residual U-Net	Validation set	0.901	0.456	0.273	0.543
Our approach		**0.952**	**0.665**	**0.656**	**0.757**
Residual U-Net	Complete training set	0.901	0.480	0.270	0.550
Our approach		**0.937**	**0.683**	**0.523**	**0.714**

Fig. 2. Visualization of predicted segmentation map and corresponding label for an image from validation set.

Table 2. Dice score for individual classes and mean across classes of our approach on test set.

	Mean sampled average Dice	Mean sampled average SD	Tumor Dice
Our approach	0.7038	0.5059	0.566

4 Conclusion

In this paper, we propose an enhancement for existing 3D U-Net model using attention-based blocks. The model we have used is a modified version of U-Net with spatial and channel attention modules. We preprocess and augment the data to improve the generalizability of our segmentation model. We compare our architecture (Mean Dice: 0.757) with Residual U-Net (Mean Dice: 0.543) architecture and show that our proposed architecture, which is inferior in depth and number of feature maps as compared to Residual U-Net, manages to outperform the Residual U-Net by significant margin on validation set after being trained for similar number of epochs. Our approach ranked 18th on the final KiTS21 challenge leaderboard. Our proposed approach has potential to improve results by further increasing the depth and number of feature maps and using a larger sized chunks to improve spatial context.

References

1. Ronneberger, O., Fischer, P., Brox, T.: U-Net: convolutional networks for biomedical image segmentation. In: Navab, N., Hornegger, J., Wells, W.M., Frangi, A.F. (eds.) MICCAI 2015. LNCS, vol. 9351, pp. 234–241. Springer, Cham (2015). https://doi.org/10.1007/978-3-319-24574-4_28
2. Kerfoot, E., Clough, J., Oksuz, I., Lee, J., King, A.P., Schnabel, J.A.: Left-ventricle quantification using residual U-Net. In: Pop, M., et al. (eds.) STACOM 2018. LNCS, vol. 11395, pp. 371–380. Springer, Cham (2019). https://doi.org/10.1007/978-3-030-12029-0_40
3. Xu, K., et al.: Show, attend and tell: Neural image caption generation with visual attention. In: ICML (2015)
4. Oktay, O., et al.: Attention U-Net: Learning where to look for the pancreas, arXiv preprint arXiv:1804.03999 (2018)
5. Hu, J., Shen, L., Sun, G.: Squeeze-and-excitation networks. In: Proceedings of the IEEE Conference on Computer Vision and Pattern Recognition, pp. 7132–7141 (2018)
6. Roy, A.G., Navab, N., Wachinger, C.: Concurrent spatial and channel 'squeeze & excitation' in fully convolutional networks. In: Frangi, A.F., Schnabel, J.A., Davatzikos, C., Alberola-López, C., Fichtinger, G. (eds.) MICCAI 2018. LNCS, vol. 11070, pp. 421–429. Springer, Cham (2018). https://doi.org/10.1007/978-3-030-00928-1_48
7. Rickmann, A.-M., Roy, A.G., Sarasua, I., Navab, N., Wachinger, C.: 'Project & excite' modules for segmentation of volumetric medical scans. In: Shen, D., et al. (eds.) MICCAI 2019. LNCS, vol. 11765, pp. 39–47. Springer, Cham (2019). https://doi.org/10.1007/978-3-030-32245-8_5
8. Heller, N., et al.: The kits19 challenge data: 300 kidney tumor cases with clinical context, ct semantic segmentations, and surgical outcomes. arXiv preprint arXiv:1904.00445 (2019)
9. Sabarinathan, D., Beham, M., Roomi, S.: Hyper Vision Net: Kidney Tumor Segmentation Using Coordinate Convolutional Layer and Attention Unit. ArXiv abs/1908.03339 (2019)
10. Zhao P, Zhang J, Fang W, Deng S. SCAU-Net: spatial-channel attention U-Net for gland segmentation. Front Bioeng. Biotechnol. 8, 670 (2020). https://doi.org/10.3389/fbioe.2020.00670
11. Ma, N., et al.: Project-MONAI/MONAI: 0.6.0. Zenodo (2021). https://doi.org/10.5281/zenodo.5083813
12. Woo, S., Park, J., Lee, J.-Y., Kweon, I.S.: CBAM: convolutional block attention module. In: Ferrari, V., Hebert, M., Sminchisescu, C., Weiss, Y. (eds.) ECCV 2018. LNCS, vol. 11211, pp. 3–19. Springer, Cham (2018). https://doi.org/10.1007/978-3-030-01234-2_1
13. Isensee, F., Maier-Hein, K.H.: An attempt at beating the 3D U-Net. arXiv preprint arXiv:1908.02182 (2019)

Transfer Learning for KiTS21 Challenge

Xi Yang[1], Jianpeng Zhang[1], Jing Zhang[2], and Yong Xia[1,3(✉)]

[1] National Engineering Laboratory for Integrated Aero-Space-Ground-Ocean Big Data Application Technology, School of Computer Science and Engineering, Northwestern Polytechnical University, Xi'an 710129, China
`james.zhang@mail.nwpu.edu.cn, yxia@nwpu.edu.cn`
[2] The University of Sydney, Sydney, NSW 2008, Australia
[3] Research & Development Institute of Northwestern Polytechnical University in Shenzhen, Shenzhen 518057, China

Abstract. Transfer learning has witnessed a recent surge of interest after proving successful in multiple applications. However, it highly relies on the quantity of annotated data. Constrained by the labor cost and expertise, it is hard to annotate sufficient organs and tumors at the voxel level for medical image segmentation. Consequently, most bench-mark datasets were collected for the segmentation of only one type of organ and/or tumor, and all task-irrelevant organs and tumors were annotated as the background. We aim to make use of these partially but plentifully labeled datasets to boost the segmentation performance of the annotation-limited KiTS21 segmentation task. To this end, we first construct a general medical image segmentation model that learns to segment these partially labeled organs or tumors. Then we transfer its pretrained weights to a specific downstream task, i.e., KiTS21. The primary experiments demonstrate the effectiveness of the proposed transfer learning strategy. Our method achieves 0.890 Dice score, 0.805 SurfaceDice, and 0.822 Tumor Dice in the KiTS21 challenge.

Keywords: Transfer learning · Limited annotation · Kidney tumor segmentation

1 Introduction

Automatic kidney tumor segmentation in computed tomography images is one of the most important tasks in the computer-aided diagnosis of kidney diseases. Although deep learning has achieved great success in many medical applications, kidney tumor segmentation remains challenging due to its limited annotations, which is a common issue for most medical image segmentation tasks.

Fortunately, there are more and more open-source benchmarks available for the development of medical image segmentation algorithms. However, most of

X. Yang and J. Zhang—Equal contribution. This work was done during Xi Yang's internship at JD Explore Academy.

them suffer from the partially labeled issue due to the intensive cost of annotations. To address this issue, Zhang *et al.* [7] proposed a dynamic on-demand network (DoDNet) that learns to segment multiple organs and tumors by using partially labeled datasets. This makes it more convenient to learn a single segmentation network from the diverse labeled datasets. In this paper, we investigate the transfer learning problem from partially labeled datasets to downstream tasks. We conduct experiments on the KiTS21 dataset. The primary results have demonstrated the effectiveness of the proposed transfer learning strategy.

2   Methods

Our Method is heavily based on DoDNet [7] and nnUNet [5], the pipeline consists of two-part: first, we use dynamic head pre-train our backbone on Multi-Organ and Tumor Segmentation (MOTS) [7] dataset, then transfer the pre-trained weight on the KiTS21 task. We illustrate the structure of our model in Fig. 1. In the downstream task, we don't use dynamic filter generating and replace the dynamic head with a convolution layer.

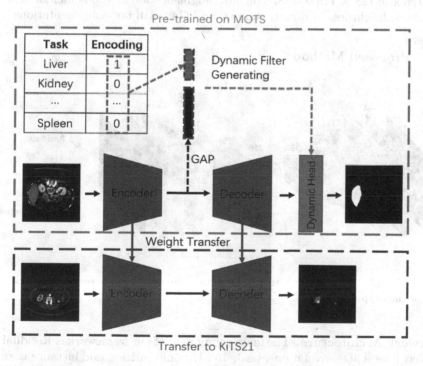

Fig. 1. The pipeline of our proposed method. We use dynamic head pre-train a segmentation network on several partially labeled datasets, and then transfer the pre-trained weights to the KiTS21 task.

2.1 Training and Validation Data

MOTS is composed of seven partially labeled sub-datasets, involving seven organ and tumor segmentation tasks (including LiTS19 [1], KiTS19 [4], and Medical Segmentation Decathlon [6]). There are 1,155 3D abdominal CT scans collected from various clinical sites around the world, including 920 scans for training and 235 for the test.

Because KiTS21 dataset contains KiTS19 dataset, so MOTS has overlapped data with KiTS21. Therefore, we only use 210 cases that have been used in MOTS pre-train for fine-tuning. According to MOTS, we choose 168 images for training and 42 for validation. Note that our final submission model is fine-tuned on all the official KiTS21 training set. In addition, we use voxel-wise majority voting (MAJ) for training and validation.

2.2 Preprocessing

Our pre-processing strategy follows nnUNet [5]. We resample all cases to a common voxel spacing of $0.78126 \times 0.78125 \times 0.78125$, and train the network with a patch size $128 \times 128 \times 128$. The data augmentation methods include scaling, rotations, brightness, contrast, gamma, and Gaussian noise augmentations.

2.3 Proposed Method

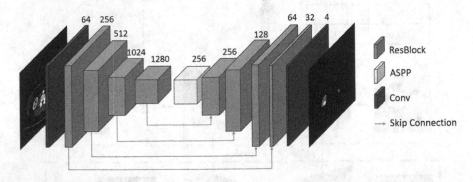

Fig. 2. Detailed network architecture. Number on the blocks represents the channel size of the outputs.

Network Architecture. The main component of our framework is Residual 3D U-Net. It uses 3D convolutions, LeakyReLU nonlinearities, and instance normalization. Upsampling is performed via transposed convolution and downsampling is performed with strided convolutions. The residual blocks of encoder are composed of Conv-instnorm-Conv-instnorm-Conv-instnorm-LeakyReLU. As shown in Fig. 2, the encoder has 4 stages. In each stage, we perform downsampling at the first residual block, then repeat these basic residual blocks (without down-sampling) 2, 3, 5, and 2 times, respectively. Inspire by [2], we use ASPP to capture objects as well as useful image context at multiple scales. Different from the encoder, the residual blocks of the decoder are composed of Conv-instnorm-LeakyReLU-Conv-instnorm, which are similar to [3]. These residual blocks are implemented in every stage of the decoder only once.

Loss Function. We train the model with the combination of dice loss and cross-entropy loss. For the two Hierarchical Evaluation Classes (HECs), i.e., **Kidney and Masses** and **Kidney Mass**, we design a HECs-based loss to optimize it. We consider HECs as the foreground and the rest as the background, and jointly use the Dice loss and binary cross-entropy loss as the objective for each task. Our total loss is a combination of the following three-part:

$l_{original}$: Computed on all classes (i.e. Kidney, Tumor and Cyst),
l_{HEC_1}: Only computed on HEC **Kidney and Masses**,
l_{HEC_2}: Only computed on HEC **Kidney Mass**.

The final loss function is:

$$l_{total} = l_{original} + \alpha l_{HEC_1} + \beta l_{HEC_2}, \tag{1}$$

where α and β are weight of each loss.

Table 1. Performance of different methods. 'SD' means Surface Dice. Kidney, masses and tumor represent HECs **Kidney and Masses**, **Kidney Mass** and **Tumor**, respectively. All methods are trained based on the nnUNet framework for 1,000 epochs with the same training strategies.

Method	Dice Kidney	Dice Masses	Dice Tumor
nnUNet	0.9718	0.8498	**0.8596**
Ours(w/o pre-train)	0.9706	0.8603	0.8522
Ours(w/ pre-train)	**0.9721**	**0.8634**	0.8577

Method	SD Kidney	SD Masses	SD Tumor
nnUNet	0.9410	0.7402	0.7443
Ours(w/o pre-train)	0.9384	0.7476	0.7358
Ours(w/ pre-train)	**0.9404**	**0.7510**	**0.7450**

Strategy. The stochastic gradient descent (SGD) algorithm with a momentum of 0.99 was adopted as the optimizer. To reduce the computational cost in the ablation experiment, all results we reported are obtained by training the models for 100 epochs based on the nnUNet framework. The learning rate was initialized to 0.01 and decayed according to a polynomial policy $lr = lr_{init} \times (1 - \frac{k}{K})^{0.9}$, where the maximum epoch K was set to 1,000.

3 Results

We use the KiTS21's official code to generate 'groups' of sampled segmentation and evaluate our predictions. The volumetric Dice coefficient and the Surface Dice are used as the evaluation metrics. The results summarized in Table 1 demonstrate the superior performance of our method over the nnUNet baseline. Some examples of our prediction results are depicted in Fig. 3.

Table 2. Ablation study of the proposed HEC-based loss functions.

Loss Settings	Dice Kidney	Dice Masses	Dice Tumor
$l_{original}$	0.9128	0.6694	0.6372
$l_{total(\alpha=1,\beta=1)}$	0.9053	0.6422	0.6297
$l_{total(\alpha=0.1,\beta=0.3)}$	**0.9237**	**0.7114**	**0.6835**

Loss Settings	SD Kidney	SD Masses	SD Tumor
$l_{original}$	0.8494	0.5072	0.4733
$l_{total(\alpha=1,\beta=1)}$	0.8336	0.4904	0.4616
$l_{total(\alpha=0.1,\beta=0.3)}$	**0.8574**	**0.5289**	**0.5045**

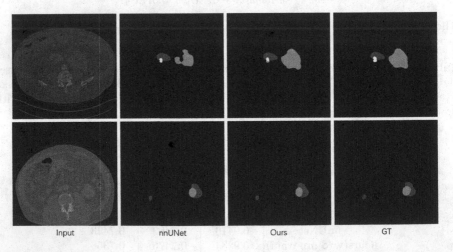

Input   nnUNet   Ours   GT

Fig. 3. Visualization of the segmentation results of case 151 (the first row) and 175 (the second row).

We use $l_{original}$ (original loss function of nnUNet) and our HEC-based loss to perform the ablation study on a basic nnUNet framework. In order to reduce the computational cost in the ablation experiment, we only train nnUNet with 100 epochs. From Table 2, we can see that $\alpha = 0.1$ and $\beta = 0.3$ achieve the best performance.

4   Discussion and Conclusion

We used dynamic filter generation and various partially labeled datasets for pre-training in this research to provide a two-stage semantic segmentation pipeline for kidney and tumor segmentation. The results of the experiments show that the suggested transfer learning approach successfully transfers pre-trained weights from MOTS to KiTS21 tasks. It shows that other small-sample 3D medical image segmentation tasks can benefit from a pre-trained 3D network.

Acknowledgment. This work was supported in part by the National Natural Science Foundation of China under Grants 61771397 and 62171377, and in part by the Science and Technology Innovation Committee of Shenzhen Municipality, China, under Grants JCYJ20180306171334997.

References

1. Bilic, P., et al.: The liver tumor segmentation benchmark (lits). arXiv preprint arXiv:1901.04056 (2019)
2. Chen, L.C., Papandreou, G., Kokkinos, I., Murphy, K., Yuille, A.L.: Deeplab: semantic image segmentation with deep convolutional nets, atrous convolution, and fully connected CRFs. IEEE Trans. Pattern Anal. Mach. Intell. **40**(4), 834–848 (2017)
3. He, K., Zhang, X., Ren, S., Sun, J.: Deep residual learning for image recognition. In: Proceedings of the IEEE conference on computer vision and pattern recognition, pp. 770–778 (2016)
4. Heller, N., et al.: The kits19 challenge data: 300 kidney tumor cases with clinical context, CT semantic segmentations, and surgical outcomes. arXiv preprint arXiv:1904.00445 (2019)
5. Isensee, F., et al.: nnU-Net: self-adapting framework for U-Net-based medical image segmentation. arXiv preprint arXiv:1809.10486 (2018)
6. Simpson, A.L., et al.: A large annotated medical image dataset for the development and evaluation of segmentation algorithms. arXiv preprint arXiv:1902.09063 (2019)
7. Zhang, J., Xie, Y., Xia, Y., Shen, C.: DoDNet: learning to segment multi-organ and tumors from multiple partially labeled datasets. In: Proceedings of the IEEE/CVF Conference on Computer Vision and Pattern Recognition, pp. 1195–1204 (2021)

Author Index

Adam, Jannes   13
Agethen, Niklas   13
Angelini, Elsa   143

Bohnsack, Robert   13
Bugata, Peter   90

Chen, Huai   53
Chen, Zhiwei   28
Chen, Zhuo   123

Dong, Zhangfu   22
Drotar, Peter   90

Finzel, René   13
Fu, Rongda   59

Gazda, Jakub   90
Gazda, Matej   90
George, Yasmeen   137
Gilboa-Solomon, Flora   103
Gohil, Sajan   151
Golts, Alex   103
Günnemann, Timo   13

Hassan, Haseeb   116, 123
He, Tian   80
Heo, Jimin   98
Hresko, David Jozef   90
Huang, Bingding   116, 123
Huang, Liqin   80
Hubacek, David   90

Khapun, Daniel   103
Kss, Bharadwaj   35

Lad, Abhi   151
Laine, Andrew F.   143
Li, Dan   123
Li, Zhaopei   71
Lin, Chaonan   59
Liu, Hanqiang   28
Liu, Zhiyang   46

Lund, Christina B.   129
Lv, Yi   40

Meine, Hans   13

Pawar, Vivek   35
Pei, Chenhao   80
Philipp, Lena   13
Plutat, Marcel   13

Rink, Markus   13

Shats, Daniel   103
Shen, Zhiqiang   1, 71
Shi, Jiacheng   22
Shoshan, Yoel   103

Thielke, Felix   13

van der Velden, Bas H. M.   129

Wang, Junchen   40
Wang, Lisheng   53
Wen, Jianhui   71
Wu, Mengran   46
Wysoczanski, Artur   143

Xia, Yong   158
Xiao, Chuda   116
Xie, Weiguo   123
Xu, Lizhan   22
Xue, Tingting   13

Yang, Hua   1
Yang, Xi   158

Zang, Mingyang   143
Zhang, Jianpeng   158
Zhang, Jing   158
Zhang, Zhen   1, 80
Zhao, Zhongchen   53
Zheng, Shaohua   1, 59, 71
Zheng, Yaoyong   71

Printed in the United States
by Baker & Taylor Publisher Services